Computational Intelligence in Urban Infrastructure

Computational Intelligence in Urban Infrastructure consolidates experiences and research results in computational intelligence and its applications in urban infrastructure. It discusses various techniques and application areas of smart urban infrastructure including topics related to smart city management. Major topics covered include smart home automation, intelligent lighting, smart human care services, intelligent transportation systems, ontologies in urban development domain, and intelligent monitoring, control, and security of critical infrastructure systems supported by case studies.

Features:

- Covers application of AI and computational intelligence techniques in urban infrastructure planning
- Discusses characteristics and features of smart urban management
- Explores relationship between smart home and smart city management
- Deliberates various smart home techniques
- Includes different case studies for supporting and analyzing various aspects of smart urban infrastructure management.

This book is aimed at researchers, graduate students, libraries in communication networks, urban and town planning, and civil engineering.

Computational Intelligence in Urban Infrastructure

Edited by
Vinod Kumar Shukla, Piyush Maheshwari,
Purushottam Sharma and Sonali Vyas

CRC Press
Taylor & Francis Group
Boca Raton London New York

CRC Press is an imprint of the
Taylor & Francis Group, an **informa** business

First edition published 2024
by CRC Press
6000 Broken Sound Parkway NW, Suite 300, Boca Raton, FL 33487–2742

and by CRC Press
4 Park Square, Milton Park, Abingdon, Oxon, OX14 4RN

CRC Press is an imprint of Taylor & Francis Group, LLC

ISBN: 978-1-032-11173-5 (hbk)
ISBN: 978-1-032-11174-2 (pbk)
ISBN: 978-1-003-21871-5 (ebk)

DOI: 10.1201/9781003218715

Typeset in Times
by Apex CoVantage, LLC

Contents

PART I Computational Intelligence for Smart Home/Smart City Management

PART II Computational Intelligence for Urban Industry

PART III   *Computational Intelligence for Intelligent Transport and Communication*

PART IV   *Computational Intelligence Supporting Urban Infrastructure*

Editor Biographies

Vinod Kumar Shukla, PhD, is currently working with Amity University, Dubai, UAE, as Associate Professor and Head of Academics in the School of Engineering Architecture Interior Design. He completed his PhD in the field of semantic web and ontology. In September 2022, he was ranked in the top 2% scientist Stanford list 2022. He is an active member of the Institute of Electrical and Electronics Engineers (IEEE) and has published many research papers in various reputed journals and conferences. He also completed the General Management Programme at the Indian Institute of Management Ahmedabad (IIM-A). He has conducted many training programs, including training employees of Delhi Transco Limited (DTL); the Indian postal department in Megdoot Bhawan, Delhi; and Directorate General Resettlement (DGR), an interservice organization functioning directly under the Ministry of Defence (DGR). He has also written case studies on "Online Retail in UAE: Driven to Grow" and "Modi's Visit to the UAE: Uncovering New Areas of Cooperation," which were hosted by the European Case Clearing House (ECCH), UK. He has written and edited many books in the field of industry 4.0 and information technology. He holds international patents in the field of information technology.

Piyush Maheshwari, PhD, is currently a professor of computer and information sciences at the British University in Dubai, UAE. During 2015–2019), he was the dean of engineering at Amity University Dubai. He is a recognized technology leader, educator and distributed systems researcher. He earned his PhD in computer science from the University of Manchester, UK, in 1990 and B.Eng./M.Eng. from IIT-Roorkee, India. He is a senior member of the IEEE.

With an overall experience of 35 years in the ICT industry and CSE academia, he brings distinguished and well-recognized thought leadership in the areas of enterprise and solutions architecture, cloud- and IoT-based software systems, system integration, and mobile-enabled applications. He has co-authored more than 130 papers in various technical journals and international conferences and holds two international patents. He has been a keynote speaker at various international forums such as ASET 2019, ICCA 2017, NGCT 2016, ICECCS 2016, Middleware 2010 and AusGrid'06, and has served on the organizing and program committees of leading conferences, reviewed many technical articles and examined several PhD theses.

Purushottam Sharma, PhD, attained his PhD in computer science in the field of temporal data mining. He is working as an associate professor in the Department of Information Technology, Amity University, Uttar Pradesh, Noida, India. He has more than 14 years of experience in research, academia and industry, and has published more than 40 research papers in SCI, ESCI, Scopus, journals indexed in Google Scholar, and reputed international conferences. He holds multiple technical patents. Presently, he is a member of various professional bodies and a reviewer for many reputed journals. He is also a Cisco Certified Instructor Trainer ITQ (Instructor Trainer Qualification). He is also certified CCNA, CCNA(R&S) Global Certificate from Cisco System USA. He has conducted many training programs including Cisco Instructor Training to various industry professionals, instructors and university faculties. He has delivered international lectures for different subjects in e-learning mode to 29 African countries using the PAN-AFRICAN network, a joint venture of the Department of Information Technology, Ministry of Communication and Information Technology, Government of India, and Amity University, Uttar Pradesh. He has published three books under student study materials for the PAN-AFRICAN project with the title *Multimedia Technologies, Accessing the WAN and LAN, Switching and Wireless* by the publisher Amity University Online.

Sonali Vyas, PhD, has been serving as an academician for around a decade. Currently, she is working as an assistant professor (senior scale) at the University of Petroleum and Energy Studies, Uttarakhand. Her research interest includes database virtualization, data mining, and big data analytics. She is a professional member of CSI, IEEE, ACM-India, IFERP, IAENG, ISOC, SCRS and IJERT. She has been awarded "Best Academician of the Year Award (Female)" in the Global Education and Corporate Leadership Awards (GECL-2018). She is an editor of *Pervasive Computing: A Networking Perspective and Future Directions* (Springer Nature) and *Smart Farming Technologies for Sustainable Agricultural Development* (IGI Global). She acted as a guest editor of a special issue of *Computer and Electrical Engineering*. She is also a guest editor of the special issue on machine learning and software systems in the *Journal of Statistics and Management Systems.*

She is also a member of the editorial and reviewer boards in many refereed national and international journals. She has also been a member of organizing committees, national advisory boards and technical program committees at many international and national conferences. She has also chaired sessions in various reputed international and national conferences. She has authored many research papers in refereed journals/conference proceedings, as well as many book chapters and books with reputed publishers. She is also professionally associated as an examiner and paper setter at CBSE, NTA, University of Rajasthan, IIS University Jaipur and Rajasthan Technical University.

Contributors

Prasoon Banerjee
Business Solution Specialist
Coforge Limited
India

Sonali P. Banerjee
Amity Business School
Amity University
Uttar Pradesh

Tulika Baranwal
University of Petroleum and Energy
    Studies
Dehradun, Uttarakhand, India

Fatima Beena
Faculty of Business and Law
De Montfort University
Dubai, UAE

Deepshikha Bhargava
University of Petroleum and Energy
    Studies
Dehradun, Uttarakhand, India

Yagya Buttan
Amity Institute of Information
    Technology
Amity University
Noida, India

Suchi Dubey
Manipal Academy of Higher Education,
    Dubai Campus
United Arab Emirates

Soumi Dutta
Institute of Engineering and
    Management
West Bengal, India

Wonda Grobbelaar
Western Caspian University
Baku, Azerbaijan

Yash Gulati
University of Petroleum and Energy
    Studies
Dehradun, Uttarakhand, India

Deepa Gupta
Amity Institute of Information
    Technology
Amity University
Noida, India

Rahul Gupta
Amity Business School
Amity University
Uttar Pradesh

Shaurya Gupta
University of Petroleum and Energy
    Studies
Dehradun, Uttarakhand, India

Sindhu Hak Gupta
Amity School of Engineering and
    Technology
Amity University
Noida, India

Sameera Ibrahim
Department of Engineering and
    Architecture
Amity University
Dubai, UAE

Deepak Jain
Shri Mata Vaishno Devi University
Jammu, India

Bakari Juma Bakari
Department of Engineering
    and Architecture
Amity University
Dubai, UAE

Raja Kunal Pandit
University of Petroleum and Energy
    Studies
Dehradun, Uttarakhand, India

Simra Nazim
IIoT Engineer at Control INFOTECH
Bangalore, Karnataka, India

Van Chien Nguyen
Thu Dau Mot University
Thu Dau Mot City
Vietnam

Rahul Nijhawan
University of Petroleum and Energy
    Studies
Dehradun, Uttarakhand, India

Lam Oanh Ha
Thu Dau Mot University
Thu Dau Mot City, Vietnam

Sunetra Saha
Amity Business School
Amity University
Uttar Pradesh

Venkat Sai Karanam
Department of Engineering
    and Architecture
Amity University
Dubai, UAE

Komal Saxena
Amity Institute of Information
    Technology
Amity University
Noida, India

Purushottam Sharma
Department of Engineering
Amity University
Noida, Uttar Pradesh,
India

Vinod Kumar Shukla
Department of Engineering
    and Architecture
Amity University
Dubai, UAE

Asmita Singh
Amity School of Engineering
    and Technology
Amity University
Noida, India

Jitendra Singh Jadon
Amity School of Engineering
    and Technology
Amity University
Noida, India

Aiman Siraj
Department of Engineering
    and Architecture
Amity University
Dubai, UAE

Abhishek Tyagi
Amity School of Engineering
    and Technology
Amity University
Noida, India

Amit Verma
Global business Studies
Dubai, UAE

Preetha V K
Department
    of Computing
University of Stirling, RAK
United Arab Emirates

Sonali Vyas
University of Petroleum and Energy
    Studies
Dehradun, Uttarakhand, India

Yashi Yashi
University of Petroleum and Energy
    Studies
Dehradun, Uttarakhand, India

Preface

This book focuses on various computational aspects related to urban infrastructure, which is also interpreted with smart city and smart home concepts. As the population grows, we need a more efficient method for planning our cities, homes and everything around us. The book helps to understand with the help of technological factors and attributes that align learners with planning their success strategies based on explicit real-life scenarios.

The book has been designed considering emerging technologies as a prime focus, such as the Internet of Things, blockchain, artificial intelligence and various wide area network–related techniques. These techniques have redefined how the growth of smart cities, smart homes and urban infrastructure has progressed. The book is a fantastic collection of the latest technologies and their applications in smart cities, smart homes, urban infrastructure, home automation, smart parking, smart transport and intelligent communication, with real-life cases related to smart city management. The book includes major technologies utilized in the urban infrastructure, such as the Internet of Things, artificial intelligence, radio frequency identification (RFID), image processing, and low-power WAN (LPWAN). The book also focuses on healthcare management and its integration with smart approaches to address the issues in this industry.

The evolution of the industry and its technological impacts have revolved around the industry point of contact and transformed business entirely. Henceforth the pivoting industry does require the latest knowledge pool of students and learners to plan and approach the new trends more scientifically. *Computational Intelligence in Urban Infrastructure* is a reference book written specifically for professionals with a focused approach towards introducing technologies related to smart cities and urban infrastructure. This book will not only act as a guide for smart city technologies; it will also explain how urban infrastructure will support your organization's strategy to revolutionize the business by imparting know-how to use it most effectively with computational techniques.

The content has been organized in four sections with the perception of smart cities, urban infrastructure, and related technology. These four sections focus on computational intelligence for smart home/smart city management, urban industry, intelligent transport and communication, and supporting urban infrastructure. The book is research oriented and will provide a common platform for all researchers in this domain, and it also covers different levels of industry and academics, which is an added advantage.

Part I

Computational Intelligence for Smart Home/Smart City Management

1 The Smart Home in a Digital Environment

Yash Gulati, Rahul Nijhawan
and Deepshikha Bhargava

CONTENTS

DOI: 10.1201/9781003218715-2

1.1   INTRODUCTION

A smart house provides an intuitive living environment for everyday convenience. All of the digital household gadgets in contemporary smart homes may be connected to the Internet to monitor, manage, and manipulate all of the electrical devices in the house. Nowadays, with the help of the Internet of Things (IoT), advanced devices and frameworks may be provided to clients anytime and from any location. A smart home's connected devices can establish a home network that can manage a variety of gadgets and digital appliances.

The digital house is predominantly powered by wired technology, with drawbacks such as lengthy installation, wiring, and poor system manageability. A wireless sensor network is now commonly utilized to construct a smart home since it is less expensive to install and facilitates system flexibility [1]. Customers choose smart homes because they offer energy savings, home security, comfort, remote health monitoring, and improved connection [2]. The rapid advancement of networking technology has given people more significant home comforts. In the smart home environment, several technologies play a key role in connecting various electrical devices to IoT applications to monitor user actions and obtain data from home devices.

Government bodies and industries endowed vast resources to provide the general public with IoT and smart home technologies. IoT is an advanced technology that can control hardware devices, transitional devices, appliances, vehicles, and sensors that helps to interact between these devices. Most smart homes support both wired and wireless transmissions.

Many home appliances, like lighting, air conditioners, cameras, smart televisions, laundry machines, and other items in a smart home can be controlled remotely by using a smartphone [3]. Smart homes show a practical approach to viewing and controlling the devices. The following are the main aims of this chapter:

- To identify the definition and motive of a smart home to understand and realize the automation procedure in a smart home;
- To review various studies on the advantages of smart homes to find the importance of living and reliability of a smart home;

- To provide an organized review of the technologies involved in smart homes to combine the importance of IoT, cloud computing, and smart city;
- To discover issues and challenges involved in smart homes and suggest solutions to the problems faced in a smart home.

1.2 THE SMART HOME IN DIGITAL WORLD

1.2.1 DEFINITION OF A SMART HOME

The smart home uses digital sensors and transmission devices to supply services with flawless transmissions [4]. Home automation is the primary technology that makes the housekeeping or house activity device controlled [5–6]. Home automation is a type of technology that can be used to control devices like cooling and heating appliances, gates with a high-security system, and latches [7–9]. A smart home reduces human intervention and provides various intelligent systems services. Furthermore, a smart home can be defined as a classic home with many home automation technologies combined to form systems that support home services by extracting the data gathered from the home.

Compared to typical homes, a smart home operates regularly, provides a comfortable area, and provides a high-quality, comfortable day-to-day existence. A smart home also improves people's lives by allowing them to reorganize their time, promoting home privacy, providing comfort, and regulating the usage of various forms of energy. Intelligence is the core goal, with all sorts of information connected inside the family of communication gadgets, domestic appliances, and security systems via family bus technology. A smart home is a living space in which a smart home system has been installed on the house's platform. For the time being, there is no uniform standard protocol since product makers limit intelligent homes. As a result, it is the responsibility of the manufacturer to conduct a system analysis and development plan. For example, the safety system will scan the face of the person entering the residence. The height of the kitchen sink may be modified, and the exhaust facilities open automatically.

Furthermore, each wall color may enhance a mood, and the temperature of the space can be adjusted automatically based on family members and emotional shifts. We will accomplish remote monitoring, intelligent lighting, and dazzling, sound, intelligent dwellings with future technological advancements. Consider how the IoT technology may be utilized to improve the look and feel of smart homes while making them more efficient and intelligent.

1.2.2 THE MAIN COMPONENTS OF A SMART HOME SYSTEM

The smart home system includes the following main points:

- *Sensors*: These gather all home data and measure house conditions. Sensors can be connected to home gadgets. A sensor's data is gathered and regularly transmitted to the smart home server by the local network.
- *Processors*: These are required for performing internal and unified actions. They can be connected to the cloud for applications that require many resources. The local processors then process the sensor's data.

- *Application programming interface* (API): These are required as a group of software parts that follow predefined parameters format to execute exterior applications. APIs can process sensor data and manage required actions.
- To execute *commands*, actuators are required within the server or other controlling devices. The device executes the command syntax, interpreting the given data to the command syntax to be executed.
- A *database* is required to save the processed data gathered from the sensors, which can also be used for data analysis. The processed data is then saved within the connected database for future use.

1.2.3 CLOUD COMPUTING AND INTERNET OF THINGS

The Internet of Things (IoT) expansion is based on the Internet and on expanding network. It has grown extensively and made communication and information exchange work simpler. Basic Internet technologies include radio frequency identification (RFID) identification, sensor networks, and machine-to-machine (M2M) communication. The three layers of IoT structure are the core technologies necessary for the most complete and well-implemented IoT implementation. Middleware, client software, remote communication, mobile Internet, and other key technologies used in the wireless sensor network gateway are all part of the IoT. The intelligent processing technology is also an integral part of security and privacy. There are various types of cloud computing available related to the smart home. It is continuously expanding the connected services established in cyberspace, to provide vigorous extension and frequently virtualized means. Besides being a network, a cloud is also a personification of cyberspace.

1.2.4 MOTIVE OF THE SMART HOME

Its goal is to reduce human involvement in the control of laboring homes and to make it easier for consumers to access smart home devices and data [2]. A smart house provides convenience, tranquility, and compact access to control industrialization. A smart home assists customers in making decisions and informs them of their options by providing them with critical information about their key factors [10].

1.3 ADVANTAGES OF SMART HOME TECHNOLOGY

1.3.1 EMPOWERMENT OF BETTER OPPORTUNITIES

Limitless chances to initiate advanced mechanisms, network them, and manage the devices privately are the features provided by the smart home [11]. Some medical managing appliances for senior citizens will be adaptable within this atmosphere. All such network provides a contributory surrounding for the latest appliances to be tried, tested, and accomplished.

1.3.2 QUALITY OF LIFE

An advantage of a smart home is the better standard of living through the ease, amenities, and facilities it offers [4] [7] [9] [12], explains the identification of smart homes better standard of living. Additionally [13], it gives brief that mental stability and managing from any place has been most popular aspects of a smart home.

1.3.3 ENERGY SAVING AND ENVIRONMENTALLY FRIENDLY

Energy-saving strategies are used in an energy-efficient and environmentally friendly house to generate significant interest in smart energy systems that monitor vitality disintegration. Therefore, the administration of vitality utilization for household automation has influenced various plans and business proposals regarding energy wastage levels [2] [14]. To save the atmosphere took many efforts; one such attempt is implementing a green atmosphere. Researchers have also focused on green vitality in the intelligent home to decrease vitality disintegration and refrigeration systems [15]. Additionally, smart homes can also enforce water conservation [16].

1.3.4 AVAILABILITY OF INACCESSIBILITY

The most appealing characteristic of a smart home is that it is available from the inaccessible, irrespective of the residence's insurance observation. Many of these options are very beneficial, mainly when inhabitants are far off from home for a holiday, for example, authenticating and checking the positions of gadgets, and confirm the doors are latched [13].

1.3.5 OBSERVATION THROUGH ANALYZER WEB

An implemented analyzer web helps to observe ventures of appliances related to older people, to handle the risk of persistent conditions and hold a fitness evaluation [17–18]. These services provides a structure that disposes functions such as smoke analyzer, gas analyzer, inaccessible observation, and advanced latches. Systems like AMIS (Advanced Monitoring and Information System) [19] collaborate with various home structures along with IoT gateways for providing assistance to the surrounding observing part [20].

1.3.6 AMBIENT ASSISTED LIVING AND AGING IN PLACE

A smart home permits a type of industrialization that keeps pace with the times in people's homes and contributes to long-term wellness. Ambient assisted living (AAL) and aging in place (AIP) are two technologies that help smart home customers. Both are adequately renowned for scientifically upholding the area inhabitants and customers with distinct requirements to encourage those who need it and give value to their security in their surroundings. Aging in place will enable an individual to stay in their own house or shifting to another place suitable for survival [21].

Conversely [22], home automation has an important role with multi-sensors in helping aged people with their living arrangements. All over this technique, the smart homes had to be enforced to be provided with several machines to uphold the services to be fastened within the house to ensure the information depository gets decent information [8]. Animated atmosphere protection can be improved by

TABLE 1.1
Smart Home Benefits

Benefit	Description
Standard of living	It gives ease, freedom, and consolation to the ordinary individual and unusual requirements, providing a standard of living.
Vitality effective and surrounding sustainable	Advance vitality helps in keeping the disposal of power lower and green. Vitality upholds keeping it sustainable.
Outlaying availability	Welfare and certainty, for inaccessible examining and administering the home.
Examining through analyzing web	It acknowledges the activities and emotional tendency to assume the requirement of an individual through advanced assuming structures.
Ambient assisted living and aging in place	Industrialization allows the elderly and others with special needs to age gracefully in their own homes while also being secure in their surroundings.

inaccessible examining automation that might be attached to a smart home. Also, the structure is joined to the sensors' cameras to get to know the state within the house by operating the advanced cloud alarm to spot any problems within the home and give urgent help to the elderly individual or inform their friends and family [23]. Table 1.1 shows the benefits of smart homes.

1.4 SMART HOME SYSTEM DESIGN BASED ON CLOUD COMPUTING

For smart homes, cloud computing is an extensive category of distributed computing that depends on the gathered substructure, constant gauging, and massive storage capacity. With this computing, a condensed machine like an advanced mobile phone would become a customer alliance, a huge information collector attached to a house industrialization machine. The smart home's storage capability has been remarkable for processing and keeping acknowledge collected from the different analyzers. It also combines data collected by modern services such as surrounding monitoring systems, closed-circuit television (CCTV), smart light systems, and energy management.

Figure 1.1 depicts the classic smart home system layout. Typically, the home gateway connects the external network to the home in various devices and services, and house owners can issue directions or log in to the external network through the gateway.

The family gateway may also simultaneously link various home gadgets and functions, which can be connected to a cloud computing center. Figure 1.2 represents the cloud computing design into the smart home system. Overall, with the help of cloud computing, the primary function of a smart home can be used:

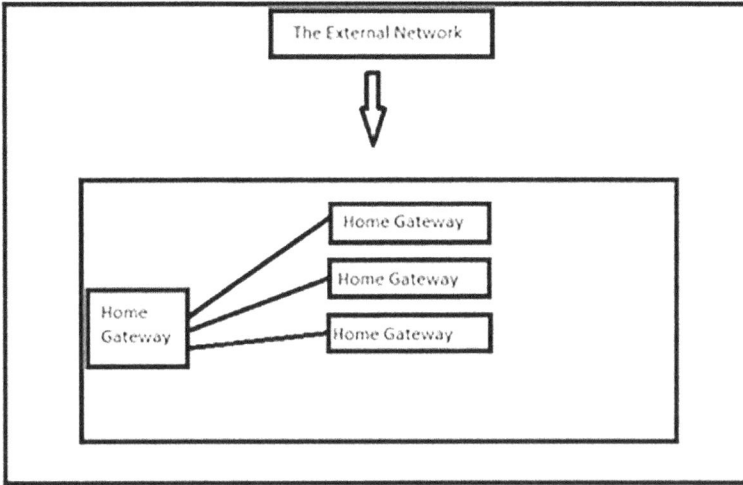

FIGURE 1.1 Traditional intelligent household structure.

- *To increase the safety of individuals who are nearby.* The smart home network can monitor things in the home through a camera and multiple detectors, as well as through the cloud of intelligent alarms to know what is going on in the house and provide remote assistance to the elderly promptly, as well as notify related families or social organizations.
- *To make life more comfortable.* Through various analyzers, cloud computing suitable for the elderly, and controlling the equipment's condition to realize the household's corresponding indicators through cloud computing.
- *To improve one's quality of life.* Smart home technology focuses on host-specific equipment, such as mobile or television remote control through the network, to manage various household gadgets. Before utilizing a single device or a small household with a range of equipment and capabilities, such as a home remote control, the owner may run the water heater, and can direct the hot tub when they reach home.

To build complex clouds for a house industrialization system, a new solution might be created by combining a bespoke home web, multiple analyzers, built-in devices, and cloud computing. The use of cloud computing to be employed in a home-industrialization system benefits the individual and the community (Figure 1.2).

1.5 EVOLUTION OF TECHNOLOGIES IN A SMART HOME

1.5.1 SMART HOME IN A SMART CITY

A smart home is one of the systems that will benefit from smart city development. To deal with rapid global urbanization, the notion of a smart city was born [24]. According to a poll conducted by the United Nations, roughly 66% of the world's

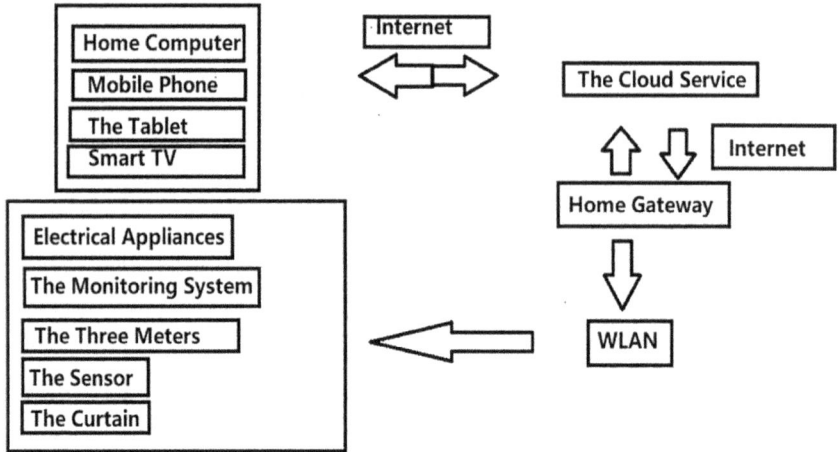

FIGURE 1.2   Cloud computing in a smart home.

population will live in cities by 2050 [25]. By 2050, it is estimated that over 6 billion people will live in cities. As a result, there is a rise in demand for food, land, sanitation, waste management, safety, and mobility [26]. Therefore, metropolitan cities are under great pressure to provide a comfortable living standard for citizens. The smart city should offer a measurable resolution to reduce the time and effort in urban development, pollution, and so forth to yield a sustainable living environment [27]. It is often a result of cities incorporating creativeness, innovation, and multi-sided ecologies that consist of numerous stakeholders with different interests collaborating to glorify a flexible environment to provide the best quality of life to the citizens [28]. Smart homes will be the best method for discovering and collecting information about people's perspectives and inducing user data with suitable privacy protection [29]. The deployment of smart cities and smart homes is being transformed by emerging technology.

1.5.2   IoT AND THE SMART HOME

IoT is the most common innovative home technology, and it may be connected to sensors, detectors, switches, and digital gadgets [30]. These gadgets may share data and resources among themselves, demonstrating intelligence. These devices can be used as a gateway from where the smart home user can be notified about the ongoing conditions within the home. These devices can be managed using a mobile phone or a computer [7]. The enhancement within the technologies of IoT integrates with intelligent frameworks within the smart home applications, allowing users to monitor their home devices and determine conditions within the home. These virtual systems are associated with functions used for mapping, aggregation, combining, and composition mechanisms that allow home applications to interoperate to perform everyday tasks of home functioning [31]. Market research revealed several technologies like IoT enable smart cities and smart homes. The IoT framework allows for a clear and faultless connection to systems and devices to complete various complicated activities. Different terms rare elated to various sensors and actuators in IoT contexts, such as smart homes. The computation of the information gathered by the sensors helps in automatic predictive systems for decision-making which provides the services required by the users [9]. Technologically, communication interfaces, computer competencies, and various protocols allow different appliances to interact and provide necessary support within the web. In general, there are a variety of home devices and appliances designed by multiple vendors that can interface with the Internet.

1.5.3 INTEGRATED WIRELESS TECHNOLOGY MANAGEMENT

Integrated wireless technology (IWT) is a type of transmission commonly used within a house or other structure to allow for inward and outward small-scale transfer via smart home automation. IWTs are frequently preferred over wired systems. Many smart home applications would be financially or otherwise restricted if wired solutions were used. On the other hand, remote improvements provide benefits such as decreased hardware and installation costs, rapid deployment, broad access, and high adaptability. Furthermore, IWT systems might be used to control and manage household equipment remotely via a graphical user interface (GUI). They also can enable communication and integration inside household energy management systems.

1.5.4 HOME ENERGY MANAGEMENT SYSTEM

The home energy management system (HEMS) was created because of the energy crisis and the effects of global warming. It has been one of the most well-known academic topics since its introduction in 1976. The HEMS allows a building's energy usage to be regulated and administered in a way that helps to reduce excessive power demand. Due to the number of hours of darkness each year, the number of HEMS installations is rising in North America in higher latitudes. In some situations, the HEMS significantly decreases total power consumption. The peak load may be reduced by 33%, and the overall cost of electricity could be reduced by 23%.

1.5.5 SMART HOME MICROCOMPUTERS

Smart home microcomputers are tiny computers that link to various equipment to automate and control the entire smart home system. They consist of a central processing unit (CPU) and its peripherals, which enable programming and looping. Their outstanding connectivity, which enables users to connect to a CPU board via various replaceable add-on modules called shields, is a crucial advantage. They enable customers to create intelligent detachable solutions with the habitat by using various expandable junctions, receiving inputs from various analyzers, and impacting the habitat's surroundings by controlling lighting or other selections.

1.5.6 HOME AUTOMATION SYSTEMS

Home automation is a smart interface that observes and learns the habits of its users, anticipating and working by their motions. Home automation has the potential to make our lives easier and more pleasant and supply some vitality ability consumption by interacting with consumers in a variety of ways. Home automation provides a chunk of the structure for smart home control. However, home automation structures should be combined with non–self-operating equipment to develop a relationship with the user. Using home automation systems exclusively, for example, would limit customers' capacity to modify their energy use. On the other hand, customers receive replies based on the management actions carried out as part of the smart home's self-operating framework.

1.5.7 ACCESS TO THE CLOUD COMPUTING CENTER

A cloud computing pivot is a group of servers solely dedicated to offering cloud computing services and obtaining data from the family gateway and storage over the web, per the defined scheme, or from the management command to compute and transmit home gateway results. Sound processing of data is needed to live in the environment of big data processing and seek out whatever information is available. Based on the analysis of the alarm system, the critical warning to recent youngsters or relevant treatment facilities, such as hospitals, police stations, and street agencies, is provided on a whole new level.

1.6 CONNECTIVITY STANDARDS IN DIGITAL ENVIRONMENTS

1.6.1 WI-FI

Wi-Fi allows computers, cell phones, and other electronic gadgets to communicate wirelessly. The IEEE Wi-Fi standard 802.11ah is the most recently created. It has now become the world's most widely used wireless technology. It supports WPA2 (Wi-Fi Protected Access) encryption and runs in the 2.4-GHz and 5-GHz frequency bands, making it easy to use. Clients with high residences cannot use the service since the embedded Wi-Fi 802.11 is only acceptable at the local entry point. IEEE 802.11ah, an ultra-low potential category of characteristics, can help by expanding the coverage area and making it easier to connect, both of which are desirable for IoT applications. Wireless network protocols used in smart home applications include Wi-Fi (IEEE 802.11), ZigBee (a low-power wireless technology), Bluetooth Low Energy, cellular, Thread and Z-wave.

1.6.2 ZIGBEE

Zigbee is a wireless communication technology with a modest data rate. It uses automated radios for personal range web based on IEEE 802.15.4 quality, with an emphasis on watching, managing and analyzing exertion. It primarily operates in the 2.4-GHz ISM band, with a 70-meter impact area. It is an open-source technology that is designed to work with a limited power supply. The Zigbee Alliance promotes the quality of home appliances, security cameras, smoke alarms, and temperature and lighting control.

1.6.3 Z-WAVE

Z-Wave is a low-potential cellular transmission technology mainly utilized for house industrialization to mitigate various exertions in the habitat and viable environment.

Across the globe, there appears to be a wide range of consumer outcomes. These features include an assembly hall, automatic meter readings, and machine-driven pool and spa maintenance.

1.6.4 BLUETOOTH LOW ENERGY

Bluetooth Low Energy (BLE) is based on IEEE 802.15.1, quality conscious of being a secure and inexpensive approach to attaching and transmitting information among many services. BLE runs in the 2.4-GHz to 2.485-GHz frequency ranges. It considerably reduces the power consumption of Bluetooth devices that can operate for a long time using a button battery. Bluetooth provides a framework of an uninterrupted attachment from smart mobile phones, enabling customers to manage house devices remotely through mobile phones. With the emphasis on IoT support, Bluetooth provides numerous enhancements like a better extended range and high data rate.

1.6.5 THREAD

This is a smartphone-based Internet protocol that works in wireless mode. This protocol uses 6LoWPAN, an open-source and unrestricted standard for building low-power private area networks (PANs) with a more significant data rate of 10–100 m, to support Ipv6. Thread is based on the physical layer 802.15.4.

1.7 CHALLENGES AND LIMITATION IN SMART HOME

Several studies were conducted in the 1980s to demonstrate the usefulness of smart homes. Digitalization and autonomous appliances have been accessible for longer, but still the smart home is a relatively new notion in many parts of the world [32]. Many inherent constraints with smart home ambitions and the crucial role of human interaction and system capabilities are some of the reasons why a smart home remains a novel concept. Various obstacles, such as technophobia, have stymied the adoption of this concept among home users. Existing protocols, network protocols, applications, devices, and other necessary technologies have failed to make smart homes profitable worldwide. Making cost-effective communication between devices for smart home settings remains a big challenge.

1.7.1 AD HOC NETWORKS HAVE A LIMITED MOBILITY RANGE

Rural residents can use smart home area networks without difficulty, despite having limited access to broadband services—for example, the architecture of a LAN that functions at 100 m changes. On the other hand, residents can create an ad hoc network using technologies like Zigbee, Z-wave, and Bluetooth to link to various smart gadgets and management systems. As a result, all small-space homes will likely embrace the new smart home. Even now, the only limitation is that network quality fluctuates due to Zigbee's 1–20 m range and requirement for network structure. Bluetooth, unlike Zigbee, has a 100 m range but uses a wireless architecture that is only ideal for short-range communication.

1.7.2   UNIVERSAL PLATFORM

In various research and related assessments, this has been proven that smart home helps to aid many older adults who reside in their homes. However, several well-designed systems have difficulty connecting to different techniques due to a lack of a distinct platform and APIs [33].

1.7.3   INTEROPERABILITY

Connectivity is considered the primary concern in home automation systems because consumers must have devices easily connected and working together [34–35]. In an advanced house, many appliances and structures come from vendors from different locations with different web affiliations. And all these appliances need to coordinate for a joint accomplishment of work. A system's ability is characterized as the capability of applications, structures and facilities to cooperate dependably in an anticipated manner. The Wi-Fi alliance's outstanding interoperability effort is responsible for Wi-Fi's enormous success. Compared to the previous version, which failed to achieve interoperability, ZigBee 3.0 will allow a larger range of appliances to connect easily, regardless of their working circumstances. A collaboration of the ZigBee Alliance and Thread Group proclaimed in April 2015 will sanction ZigBee goods to utilize the Thread agreement to ensure interoperability outcomes. In order to achieve productive interoperability, as well as other smart home automotive behaviors, numerous studies have been conducted. In a home automation system, it is challenging to adapt the outcomes of adding more devices or sensors to the current ambience. It may or may not be fruitful to the users as some of these protocols can vanish [36]. Regarding the wired sensors like environment sensors, there are some concerns too, like environment-dependent and short heterogeneous functionalities.

1.7.4   SELF-MANAGEMENT

Many intelligent devices can monitor their operating health and notify of possible issues before collapse. A few applications based on sensor networks can work without infrastructure or mobility for maintenance. The most common need for sensor nodes is for them to be self-managed, which implies that they must be completely independent of human intervention to regulate and interconnect with other devices and respond to environmental changes. The widely used technology is Z-wave, as its installation and preparation are easy with automatic address assignment, making network management more effortless. It also has an anti-interference property supported by a random back-off algorithm and escapes collisions. Zigbee is a technology with solid and self-formed computed networking and is also adaptable in connecting with several topologies [37].

1.7.5   EXTRACTED DATA'S RELEVANCE

Implementing primary data in information-driven strategies that use data processing methods or other device-experience ways to capture a good data collection requires knowledge [38]. The most critical issue noticed was the restation of the information gathered from the analyzer. It will only be pertinent once the data get filtered. The studies show that the classification and knowledge from various related sets produce widely applicable data. The more significant part of the factor is assigned to the active type [11].

1.7.6 Commercialization

Unobtainable technology or delay in commercialization are the factors that hinder smart homes from being well-liked. Most smart home commodities are brought to market before being tested well by the customers [11]. Learning shows that machinery inventors have also ignored the necessary ideas of house customers, like who the users are and how they will consume the smart home machinery, in a hurry to capitalize the commodities to the emporium.

1.7.7 Failure to Faulty Systems

Adjusted house structures [39–40] would have a replica deliberated to manifest the system's operation to reciprocate to defective structures. Thus, a self-curing structure ought to be created mandatory. In the investigation [41], physical components were suggested as an important part of self-development neighbor unearthing, intermediate entry management, provincial, and alley initiations.

1.7.8 Security and Privacy

The house industrialization structure usually has problems with system spamming and menace entirely on the insurance of the server side [42]. Moreover, administering the home industrialization structure out of the way could be a provocation regarding confidentiality and privacy [43]. Together with this, practical environment-friendly components are concerned [10]. Typically, server-side safety is minimal as no determining methodology for corroboration could give a place to ensure home industrialization structure. All these problems happen due to the utilization of web connection provisions [44].

1.7.9 Cost

In an adjustment of machinery, price plays a vital part in getting or maintaining that machinery. Machinery recognition investigation demonstrated that there had been issues concerning buying, executing, utilizing and maintenance costs [45].

1.7.10 Changes in Society

Home automation has reaped the benefits of comprehensive and subsequent civil coinage, causing unintended consequences such as more excellent mechanization reliability. It may also result in weight gain and sedentary behavior. It is a desire for integrating and equipping; for example, to analyze society drift and behaviors, we have used data collected from the whole smart home society [46].

1.7.11 MAINTAINABILITY

Maintainability is a requirement in the web environment that cogitates how trustworthy and long-running smart home network will work. The commute is done everywhere, comprising the home surrounding with a weakening fork accumulator and a new chore. As a result, the smart home web's substructure must analyze its power and location to adjust functional boundaries or choose between various services, such as decreasing standards when vitality resources are limited. The smart house web should be designed for easy maintenance, allowing faster and more cost-effective repairs of various equipment and communication components [47].

1.7.12 BANDWIDTH

Bandwidth utilization is one of the other challenge for IoT attachment. Handling in-home web also becomes difficult. With the increment figure of offers and house appliances, a large quantity of information has been discovered; as an outcome, bandwidth necessity is very usual in the advanced home. The clip field must primarily serve as the enactment, requiring a significant bandwidth. It poses a significant server problem, requiring a highly scalable server to handle all data. For this reason, a wobbly web safely delivers data between appliances and servers. A tablet might be a high-power, high-bandwidth device, but BLE offers more data capacity than Zigbee and Z-Wave.

1.7.13 POWER CONSUMPTION

Appliances attached to the IoT smart homes can give the top time to work, providing huge organization in power utilization. Many IoT appliances in the smart home transfer gestures and information in a circular chronometer to levy potentiality and CPU utilization. An organized IoT web requires little accumulator loss and also less potential utilization in such transmission. Potential utilization is elevated in quality tablets. However, Bluetooth without vitality is the best in potential utilization and has limitations in the signal area and the numeral of appliances. Large customers who can run for many years on the tiniest of accumulators are transported by organizations and appliances working together.

1.8   INFRASTRUCTURE RELATED TO THE SMART HOME

Wi-Fi is a significant component of the current smart home attachment strategy. PCs and small appliances primarily power it; Wi-Fi generally has a high information rate and requires extra power. Another alternative for linked appliances is ZigBee. Electric meters, indicators, electric lighting, and electric car charging stations are just a few of the devices that employ ZigBee. Thread is a broadcast protocol essential for mesh networking and network authentication. The most advanced consideration regarding attachment is to choose an online and leading concord that supplies extendable, adaptable, and administrative possibilities. For accessing network-based amenities, constrained application protocol (COAP) is employed to give a dependable and flexible method to the inaccessibility-produced analyzer and online information [48]. Bandwidth sent to a home or office is useless unless it can be distributed to consumer equipment. A house web and network administrator's job is to distribute bandwidth and managing connected equipment.

This structure comprises several advanced forks in a dwelling area web that several analyzers have absorbed. The automation and equipment used are attached and analyze light, solar, temperature, and humidity. An arc appliance or a home control unit (HCU) with any broadcast quality is connected to an analyzer fork. The arc appliances serve as a conduit for data transfer from a broadcast analyzer web to the Internet. Understandably, context repetition becomes unavoidable when much data is collected. As a result, there is a need for a beneficial purification of factor strainer design. One of the most notable disadvantages was that the information received from the analyzer contains many reiterations, and further analysis is only done after file refinement.

1.9 SMART HOME–RELATED SOLUTIONS

It is apprised that a laboratory surrounding will resolve the non-predictable problems of loading recent appliances or analyzers into the present sets of any advanced home [35]. It combines the benefits of power line transmission (PLC) with the ability of cellular transmission to create a PLC system with evaluated speed, infrastructure-independent omnipresent house ingress, and good bandwidth organization for connecting stable and phone appliances in a house [49]. Furthermore, artificial intelligence (AI) innovation can construct to meet consumers' needs. Using multiple factors, AI data removes superfluous information and gives filtered value to the information elimination for a more accommodating facility. Cloud automation is likewise a standard program, and it is a likely proposal to the vast information capacity [46].

1.10 CONCLUSION

The emergence of "smart homes" has highlighted the need for home devices to interact with all appliances to provide the home customer with the ability to examine or manage their household work and raise the standard of living. As automation progresses, many automated advanced recommendations for house clients are being catered to. Along with the automation, the actual end users must be carefully considered. Ethics, price, automation acceptability, safety, mechanical features, legal, organizational structure, and constraints must be thoroughly explored to better execute the smart house in the future.

Individuals seeking a high standard of life have increasingly been drawn to smart homes. The IoT will soon connect even the most basic household equipment to the Internet. The accessible quality in the emporium, which is the growing slab of the home web and invention, is also summarized in this chapter. In addition, the smart home web's provocation had to rely on its connectivity. Zigbee components are used for lower-level bandwidth exertions, whereas Wi-Fi is used for higher-level bandwidth exertions. As a result, a high degree of quality among protocols is expected in a smart home web. A quality-based architecture that supports integration across many transmission protocols is now being recommended for the advanced operation of a smart home.

REFERENCES

[1]. Li, M., Gu, W., Chen, W., He, Y., Wu, Y., and Zhang, Y. 2018. Smart home: Architecture, technologies, and systems. Procedia Computer Science, 131, 393–400. https://doi.org/10.1016/j.procs.2018.04.219.

[2]. Yang, H., Lee, H., and Zo, H. 2017. User acceptance of smart home services: An extension of the theory of planned behavior. Industrial Management & Data Systems, 117(1), 68–89.

[3]. Global Smart Home Market to Exceed $53.45 Billion by 2022: Zion Market Research. https://globenewswire.com/newsrelease/2018/01/03/1281338/0/en/Global SmartHomeMarket-to-Exceed-53-45-Billion-by-2022-Zion-MarketResearch.html. Accessed: 2018–6–6.

[4]. Babakura, A., Sulaiman, M.N., Mustapha, N., and Perumal, T. 2014. Home-based decision model for the smart home environment. International Journal of Smart Home, 8(1), 129–138.

[5]. Wang, M., Zhang, G., Zhang, C., Zhang, J., and Li, C. 2013, June. An IoT-based appliance control system for smart homes. In Intelligent Control and Information Processing (ICICIP), 2013 Fourth International Conference on, pp. 744–747. IEEE.

[6]. Zaidi, S.F.N., Shukla, V.K., Mishra, V.P., and Singh, B. 2021. Redefining home automation through voice recognition system. In Emerging Technologies in Data Mining and Information Security (pp. 155–165). Springer, Singapore.

[7]. Vani, K.S., and Shrinidhi, P.C. 2015. "Automatic tap control system in the smart home using Android and Arduino. International Journal of Computer Applications, 127, 8.

[8]. Mano, L.Y., Faiçal, B.S., Nakamura, L.H., Gomes, P.H., Libralon, G.L., Meneguete, R.I., Geraldo Filho, P.R., Giancristofaro, G.T., Pessin, G., Krishnamachari, B., and Ueyama, J. 2016. Exploiting IoT technologies for enhancing health smart homes through patient identification and emotion recognition. Computer Communications, 89, 178–190.

[9]. Feng, S., Setoodeh, P., and Haykin, S. 2017. Smart home: Cognitive interactive people-centric Internet of Things. IEEE Communications Magazine, 55(2), 34–39.

[10]. Wilson, C., Hargreaves, T., and Hauxwell-Baldwin, R. 2015. Smart homes and their users: A systematic analysis and key challenges. Personal and Ubiquitous Computing, 19(2), 463–476.

[11]. Shukla, V.K., and Singh, B. 2019. Conceptual framework of smart device for smart home management based on RFID and IoT. Amity International Conference on Artificial Intelligence (AICAI), 2019, 787–791. https://doi.org/10.1109/AICAI.2019.8701301.

[12]. Ni, Q., García Hernando, A.B., and Pau de la Cruz, I. 2016. A context-aware system infrastructure for monitoring activities of daily living in a smart home. Journal of Sensors. doi:10.1155/2016/9493047

[13]. Brush, A.J., Lee, B., Mahajan, R., Agarwal, S., Saroiu, S., and Dixon, C. 2011. May. Home automation in the wild: challenges and opportunities. In Proceedings of the SIGCHI Conference on Human Factors in Computing Systems (pp. 2115–2124). ACM, New York.

[14]. Athreya, A. P., and Tague, P. 2013. June. Network self-organization in the Internet of Things. In 2013 IEEE International Workshop of Internet-of-Things Networking and Control (IoT-NC), pp. 25–33. IEEE, New Orleans, Lousiana, USA, 24 June, 2013.

[15]. Komal, S., Basit, A., and Vinod Kumar, S. 2021. Green internet of things (G-IoT) technologies, application, and future challenges. Green Internet of Things and Machine Learning: Towards a Smart Sustainable World, 317–348.

[16]. Tripathi, G., Singh, D., and Jara, A.J. 2014. A survey of internet-of-things: Future vision, architecture, challenges, and service. In IEEE World Forum on Internet of Things (WFIoT), 287–292.

[17]. Kim, H.C. 2015. Acceptability engineering: the research of user acceptance of innovative technologies. Journal of Applied Research and Technology, 13(2), 230–237.

[18]. Yang, R., and Newman, M.W. 2013. Learning from a learning thermostat: Lessons for intelligent systems for the home. Proceedings of the 2013 ACM International Joint Conference on Pervasive and Ubiquitous Computing (pp. 93–102). ACM, New York.

[19]. Skubic, M., Guevara, R.D., and Rantz, M. 2015. Automated health alerts using in-home sensor data for embedded health assessment. IEEE Journal of Translational Engineering in Health and Medicine, 3, 1–11.

[20]. Azimi, I., Rahmani, A.M., Liljeberg, P., and Tenhunen, H. 2017. Internet of things for remote elderly monitoring: research from a user-centered perspective. Journal of Ambient Intelligence and Humanized Computing, 8(2), 273–289.

[21]. Fong, B.Y.F., and Law, V.T. 2018. Aging in place. In Sustainable Health and Long-Term Care Solutions for an Aging Population, Hersey; IGI Global, 259–276.

[22]. Ranasinghe, S., Al Machot, F., and Mayr, H.C. 2016. A review on applications of activity recognition systems with regard to performance and evaluation. International Journal of Distributed Sensor Networks, 12(8), 1550147716665520.

[23]. Lin, Y. 2015. December. Study of smart home system based on cloud computing and the key technologies. In 2015 International Conference on Computational Intelligence and Communication Networks (CICN), pp. 968–972. IEEE, GGITS, Jabalpur, India, Dec, 12–14, 2015.

[24]. Macke, J., Casagrande, R.M., Sarate, J.A.R., and Silva, K.A. 2018. Smart city and quality of Life: Citizens' perception in Brazilian case research. Journal of Cleaner Production, 182, 717–726.

[25]. Shirehjini, A.A.N., and Semsar, A. 2017. Human interaction with IoT-based smart environments. Multimedia Tools and Applications, 76(11), 13343–13365.

[26]. Hansain, S., Gaur, D., and Shukla, V.K. 2021. Impact of emerging technologies on future mobility in smart cities by 2030. 2021 9th International Conference on Reliability, Infocom Technologies and Optimization (Trends and Future Directions) (ICRITO), 1–8. https://doi.org/10.1109/ICRITO51393.2021.9596095.

[27]. UNEP. 2016. The Emissions Gap Report 2016. United Nations Environment Programme (UNEP), Nairobi.

[28]. Capdevila, I., and Zarlenga, M.I. 2015. Smart city or smart citizens? The Barcelona case. Journal of Strategy and Management, 8(3), 266–282. https://doi.org/10.1108/JSMA032015-0030.

[29]. Gaur, A., Scotney, B., Parr, G., and McClean, S. 2015. Smart city architecture and its applications based on IoT. Procedia Computer Science, 52, 1089–1094.

[30]. Wu, C.-L., Tseng, Y.-S., and Fu, L.-C. 2013. Spatio-temporal feature enhanced semi-supervised adaptation for activity recognition in IoT-based context-aware smart homes. In 2013 IEEE International Conference on Green Computing and Communications and IEEE Internet of Things and IEEE Cyber, Physical and Social Computing, IEEE, Beijing, China, pp. 460–467. doi: 10.1109/GreenCom-iThings-CPSCom.2013.94.

[31]. Tao, M., Zuo, J., Liu, Z., Castiglione, A., and Palmieri, F. 2018. Multi-layer cloud architectural model and ontology-based security service framework for IoT-based smart homes. Future Generation Computer Systems, 78, 1040–1051.

[32]. U.N. (United Nations). The state of African cities 2014: UN-habitat. United Nations Human Settlements Programme (UN-Habitat) (Nairobi: United Nations Human Settlements Programme, 2014).

[33]. Fattah, S.M.M., Sung, N.M., Ahn, I.Y., Ryu, M., and Yun, J. 2017. Building IoT services for aging in place using standard-based IoT platforms and heterogeneous IoT products. Sensors, 17(10), 2311.

[34]. Samuel, S.S.I. 2016. March. A review of connectivity challenges in IoT-smart home. In 2016 3rd MEC International Conference on Big Data and Smart City (ICBDSC), pp. 1–4. IEEE, Muscat, Oman 15–16 March 2016.

[35]. Miori, V., and Russo, D. 2014. Domotic evolution towards the IoT. In Advanced Information Networking and Applications Workshops (WAINA), 2014 28th International Conference (pp. 809–814). IEEE, Piscataway.

[36]. Mennicken, S., Vermeulen, J., and Huang, E.M. 2014. From today's augmented home to tomorrow's smart homes: new directions for home automation research. In Proceedings of the 2014 ACM International Joint Conference on Pervasive and Ubiquitous Computing (pp. 105–115). ACM, New York.

[37]. Shukla, V.K., and Singh, B. 2019. Conceptual framework of smart device for smart home management based on RFID and IoT. 2019 Amity International Conference on Artificial Intelligence (AICAI), 2019, 787–791. https://doi.org/10.1109/AICAI.2019.8701301.

[38]. Bouchard, K., and Giroux, S. 2015. July. Smart homes and the challenges of data. In Proceedings of the 8th ACM International Conference on Pervasive Technologies Related to Assistive Environments, pp. 1–4.

[39]. Yang, C, Mistretta, E, Chaychian, S., and Siau, J. 2016. Smart Home System Network Architecture, Paper presented at 1st EAI International Conference on Smart Grid Inspired Future Technologies, Liverpool, United Kingdom.

[40]. Strangers, Y. 2016. Envisioning the smart home: Reimagining a smart energy future. Digital Materialities: Design and Anthropology, 61–76.

[41]. Cook, D.J. 2012. How smart is your home? Science, 335(6076), 1579–1581.

[42]. Gabhane, M.J.P., Thakare, M.S., and Craig, M.M. 2017. Smart homes system using internet-of-things: Issues, solutions, and recent research directions. International Research Journal of Engineering and Technology (IRJET), 4(5), 1965–1969.

[43]. Almusaylim, Z.A., and Zaman, N. 2018. A review on smart home present state and challenges: linked to context-awareness Internet of things (IoT). Wireless Networks, 1–12.

[44]. Fernandez-C, M., and Bellotto, N. 2016. On-line inference comparison with Markov Logic Network engines for activity recognition in AAL environments. Intelligent Environments (I.E.), 2016 12th International Conference, 136–143.

[45]. Park, E., Kim, S., Kim, Y., and Kwon, S.J. 2018. Smart home services as the next mainstream of the ICT industry: Determinants of the adoption of smart home services. Universal Access in the Information Society, 17(1), 175–190.

[46]. Guiry, J.J., Van de Ven, P., and Nelson, J. 2014. Multi-sensor fusion for enhanced contextual awareness of everyday activities with ubiquitous devices. Sensors, 14(3), 56875701.

[47]. Mckeown, A., Rashvand, H., Wilcox, T., and Thomas, P. 2015, August. Priority SDN controlled integrated wireless and powerline wired for smart-home internet of things. UIC/ATC/ScalCom, pp. 1825–1830.

[48]. Mendes, T.D., Godina, R., Rodrigues, E.M., Matias, J.C., and Catalão, J.P. 2015. Smart home communication technologies and applications: Wireless protocol assessment for home area network resources. Energies, 8(7), 7279–7311.

[49]. Ibrahim, S.V., Shukla, K., and Bathla, R. 2020. Security enhancement in smart home management through multimodal biometric and passcode. International Conference on Intelligent Engineering and Management (ICIEM), 420–424. https://doi.org/10.1109/ICIEM48762.2020.9160331.

2 Adoption of the Internet of Things in the Smart Home

Tulika Baranwal, Rahul Nijhawan and Sonali Vyas

CONTENTS

2.1 INTRODUCTION

The phrase "smart home" refers to a current application of ubiquitous computing that integrates expertise into the governance and administration of homes for "peace, wellness, safety, privacy, and energy-saving" [1]. This point of view, which was expressed in this study, exemplifies the convergence of two complementary perspectives on intelligent home capabilities: specific user and system, which focus on occupant satisfaction and building system efficiency, respectively [2]. The user- or residential-centered approach to smart buildings has evolved through time. It dates back to the first half of the twentieth century. The system- or building-centric approach emerged mostly with the advancement of information and communication technology (ICT) and the installation of smart power generation.

Over the last few years, there has been a surge in the development of new "smart" gadgets which can connect to the Internet and be remotely controlled via apps. The Internet of Things (IoT) is a network of devices and items having detectors, circuits, applications, and web access. It has resulted in creating a virtualized IoT-based smart home software development company that incorporates another upcoming cloud computing technology [3].

The IoT technology moves the focus more on connectivity and informative judg-ment, implying that if a device is linked to other devices, it may gain value. On the other extreme, the Internet of Things is more than a gathering of gadgets and sensing devices connected by a wired or wireless network; it is a computed value of the phys-ical and virtual worlds, where people and objects interact. It may be viewed as an interconnected platform that connects various sized networking [4] systems to create a giant global network.

The new aspects of smart homes that have evolved as a repercussion of the adop-tion of novel computer paradigms are examined in this chapter. The notion of a smart home and its cloud-based architecture are presented in section 2. The link between smart households and smart grids is also briefly mentioned. Section 3 highlights dif-ferent types of software that are essential for home automation based on the Internet of Things: operating systems, occupant tracking, and data collection and process-ing software. Section 4 delves into the most extensively used wired and wireless communication protocols and technologies. Section 5 focuses on the issues about the confidentiality of the information and data privacy in smart home devices. The future trends and challenges in smart house development are discussed in section 6, followed by a conclusion of the research in section 7.

2.2   AN OVERVIEW OF THE SMART HOME

This section explains about the history of home automation techniques. It begins with concepts of smart buildings and intelligence within smart homes, then goes into virtualized home automation architecture. It also underlines key characteristics of smart homes that render these useful, willing members in today's intelligent grids.

2.2.1 Smart Homes

The intelligent home was first proposed in the 1930s as a marvel of household efficiency in "homes of the future" fantasies [5]. The majority of hopes for "new heights of richness, pleasure, and indulgence," as well as "benefits of contemporary living with less effort from householders," did not materialize until the century's last decades. Meanwhile, the technology on household performance has switched to fuel productivity [2]. Darby separates both pointers of home automation concepts: an intelligent house is a fully labor-saving, integrated populous structure that stresses advanced techniques, ease, and (household) sustainability.

Architecture and framework research aims to build performance parameters, auxiliary facilities, decentralized power generation, and how these concerns might be addressed with the use of ICT infrastructure. One point that should be emphasized is that both definitions highlight the importance of communications in connecting equipment, enabling remote control and access, and providing services. It is important to explore how the terms "smart" and "intelligent" might be defined when discussing smart home. What distinguishes a smart house from the traditional homes where most of us still reside?

As per Edwards and Grinter, intelligence consists of four properties in light of intelligent surroundings and omnipresent computation [6]:

1. Surroundings may utilize sensor data to analyze the present status (e.g., if a motion-sensing detector is engaged, it indicates somebody is walking close).
2. The surroundings may infer its current condition by assessing a number of factors simultaneously (e.g., if there are numerous individuals at the table, the framework may deduce that it is mealtime).
3. The surroundings can forecast a client's purpose by analyzing circumstances beyond its perspective (e.g., if consecutive motion detecting devices are activated, it implies the client is going down a corridor, and the user may wish to build a path illuminated).
4. Based on the purpose assumption, the environment may respond pre-emptively (e.g., the framework may conclude to switch on the lights ahead of time, allowing the client to travel securely on the track).

An intelligent environment may be thought of as an intelligent agent that can use sensors to analyze the status of a house, its inhabitants, and their physical elements and then act on the surroundings using effectors to maximize specified performance measures. The performance metrics to be increased in a smart house [7] might be the occupant's ease and efficiency, while those to be reduced are the operation expenses [8] (i.e., the expense of fuel and other needs).

To be smart, structures ought to be flexible, context-sensitive, adaptable and predictable [9–10]. A smart dwelling area is decided via the surroundings' capacity to evolve to the desires of its occupants [11]. As a result, we need to distinguish between smart domestic automation and smart domestic, related to service, design, installation and uses. People have continually used automation technology; however, automation is the handiest part of smart surroundings (thermostats, washing machines, etc.). On the other hand, automation is liable for only some variables, so you cannot see the whole picture. On the other hand, intelligence may be defined as being capable of higher adaptation and expectation primarily based on the contemporary environment of the machine and the preceding information of its occupants, because it has a broader view of the complete surroundings via all networked gadgets and sensors.

2.2.2 ARCHITECTURE OF SMART HOMES

Some devices that are not always smart on their own are included in a smart home, including detectors, appliances, or controllers. Sensors generate facts but have little effect on the home's environment. If the owner needs to manually change the temperature based on outdoor temperature, humidity and other variables, the thermostat will not be normal. You can maintain a constant temperature, but automation is the end result, not intelligence. A smart environment can only be created by collecting and analyzing all the facts to find a style and select without user input.

Soliman [12] and his colleagues designed an intelligent house system. Sensors and actuators are used to gather information about the domain, perform particular actions, and are connected to a microcontroller. The cloud system comprises the server-site app, memory, and a front-end implementation. People can use a web application to observe their surroundings and control their devices.

Cook et al. [13] proposed a physical hierarchy that encompasses actuators (servers), detectors (motion sensors), as well as actuators (sensors of interest). Wi-Fi Talk technology and Wi-Fi ZigBee grid are used at the talk level. At the middleware level, the publish or subscribe sample is used. Jie et al. [14] evolved an organizational paradigm to solve the scalability problem. Persistent versions, as suggested, make it easy to add or exclude gadgets from your smart backend infrastructure. The structure is divided into following five levels.

1. Asset level (stationary devices, detectors and equipment);
2. Software interfaces (interface hardware, detectors, hardware);
3. Agency level (sellers are responsible for tracking individual assets using RFID tags);
4. Core level (agent supervision and main controller);
5. Personal software layer (consumer gadget management platform).

CloudThings is a virtualized framework designed by Zhou et al. [15] to quicken future IoT creation and implementation. Peripherals (Things) employ the CoAP protocol and 6LowPAN to browse the web directly. CloudThings is a web-based platform for the development, deployment, operation and creation of applications and services across the entire application infrastructure. This framework consists of following three parts.

1. *Infrastructure as a Service*: This solves all cloud infrastructure issues without worrying about computing power, memory, scalability, or server management.
2. *Platform as a Service*: This defines a framework within which development teams can design and implement functionality for things.
3. *Software as a Service*: This acts as an environment for the development and maintenance of features.

The architecture of a cloud-based smart home is depicted in Figure 2.1. The internal network comprises end systems, detectors, utilities and motors. Such gadgets interact with a portal at the network's edge, which links private systems and the outside world. End devices, detectors, computers and the cloud are all connected through an access point. In their study, Hosek et al. [16] go to hardware and technical requirements for smart indoor gateways. Guoqiang et al. [17] advocate a careful

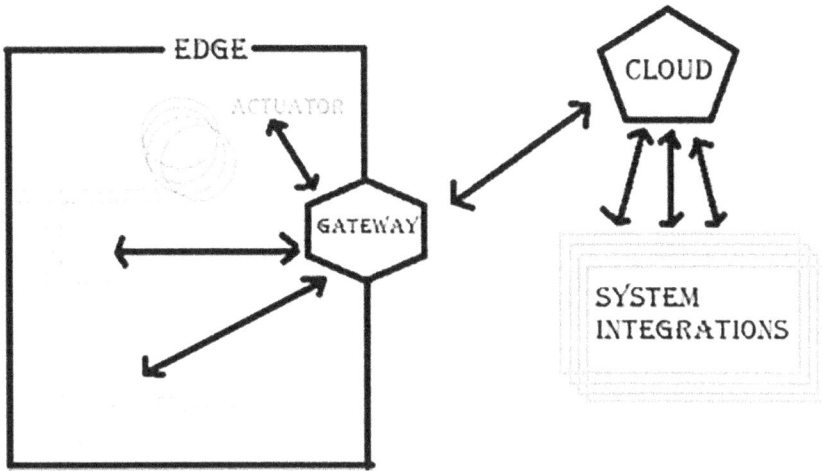

FIGURE 2.1 Cloud-based architecture of a smart home.

design of internal gateways that can translate numerous sensor statistics directly into a wide range of layouts and be applied in the form of connectivity protocols.

Essentially, a gateway is a tool that stores several dialogue structures that can be used with exit structures. It is enough to study the states from the outer edge of the community before sending them to the cloud. By connecting a smoking cessation device to the outside world, the gateway grants all other safety credentials to the smart home community, creating a smoking cessation structure (resources limited, undernourished, less secure and more vulnerable to attack). Leveraging the extra features of laptops and scalable architectures can be a threat that increases uptime, reliability, and privacy. The cloud can be used to bundle various services such as visualization technology, home automation management, administrative access and task management.

2.2.2.1 Cloud Computing

A model for providing ubiquitous, convenient, on-demand access to a common pool of configurable computing resources (e.g., networks, servers, storage, applications and services) that can be rapidly provisioned and released with minimal management effort or service provider interaction [18]. In smart home design, the cloud is utilized to gather and analyze data from a variety of sources. The cloud is the maximum advanced architecture level [19], with massive data storage and processing architecture that is exceptionally stable, adaptive and self-contained. Because of its central position in this approach, the cloud is frequently referred to as virtualized or web-based architecture [20].

2.2.2.2 Defining Fog Computing

Fog computing is a brand-new computing paradigm for superior cloud architectures invented by Cisco in 2014 [21]. It has been deployed to numerous applications [22–24]. Fog computing has positive protection and privacy concerns [25–26].

The cloud computing paradigm has flourished since its inception a decade ago. Certain demanding situations with this paradigm are beginning to be seen because the variety of IoT gadgets that want to generate extra statistics and join and transmit these statistics to the cloud increases. One of the maximum severe issues in contemporary computing is that the quantity of memory and records switch bandwidth can't maintain with the boom within the processing electricity of the CPU [27]. As a result, community bandwidth has become a bottleneck, making cloud answers inefficient. Abandoned computing solves this trouble by transferring processing towards the record source. This technique lets records be analyzed between statistics gadgets and the cloud, then transmitted as low-noise cloud records. Fog computing refers to records calculations accomplished on gateway gadgets in the smart home. The most important desires of fog computing are to decrease the number of facts dispatched to the cloud, lessen community latency, and grow gadget reaction time in question situations [28].

Another gain of fog computing is vulnerability mitigation. For example, while the net connection is lost, the smart domestic gateway keeps accumulating facts from the give up gadget after the community is created, examining it, and shipping it to the cloud. There are also two-manner blessings of using a portal tool with the cloud. Portal equipment now no longer best reduces the number of facts dispatched to the cloud but additionally lets in the cloud to ship messages to the portal, which proclaims the message to the proper give up gadget.

2.2.2.3 Edge Computing

Edge computing is another computational method. It pursues similar goals like fog computing, but differently. Lopez et al. [29] addressed the notion of edge computing, emphasizing user apps in which individuals may remain anonymous and private. Shi et al. [30] looked at several case studies to analyze the paradigm's flaws and possibilities. Shi and Dustdar [31] discovered several unsolved privacy and security issues in edge computing. As opposed to fog computing, edge computing takes localized computation a step further by allowing each smart home end device to choose whether the information shall be kept and processed locally or transferred to the cloud. The system as a whole is more decentralized, minimizing the peril of each point of negligence because each end device may make some inference and connect with the cloud. To put it another way, edge computing moves data analysis and judgment to the portals at the periphery.

2.2.3 SMART HOME WITH SMART GRID

Over the last ten years, there's been a rise in the awareness of power utilization on a global level, notably in the areas of alternative energy sources. The introduction of the smart grid concept as a vehicle for energy system upgrading has made this possible. Ten requirements were stated in the first official description of smart grid architecture [32]. Half of these characteristics are linked to home automation, including the following:

- Dynamic pricing, requirement resources, and conventional energy sources are all in the works.

- "Innovative" technology is being used for metering, grid communications, and distribution automation (real-time, automated, participatory technologies that optimize the physical operation of appliances and consumer gadgets).
- Consumer electronics and "smart" goods are being used.
- Improved power storage and apex technologies are being implemented and integrated, such as connectors in electric-based automobiles.
- Timely information and control options are being made available.

As a result, sophisticated networking capabilities bypass distribution infrastructure to enable efficient home operation. By allowing energy and records to travel through all aspects between the home and the power grid, electrical devices can become more adaptive, more responsive and deliver real-time feedback. The smart grid's response call (D.R.) is an essential component [33]. When coordinating low-priority family devices, D.R. Plus users get the ideal electric recalculation for the lowest possible price (including a washing machine or water heater). There are two types of power supply schedules:

1. Energy support management (which affects when the amount of power redistribution made up of the connected grid is used while household photovoltaic panel units are in use);
2. Power consumption management (the goal is to optimize the turn-on time of various appliances, considering the hurdles consisting of either the amount of electricity saved at the regional level or the dominant payment of electricity within the region; time of use pricing scheme).

Fuselli et al. [34] proposed a probabilistic dynamic programming approach with motion structures to improve the planning of useful resources. It is made up of individual neural networks: a community of movements that define the state being manipulated and a community of critics who evaluate the community of movements and refine the former ideal state over time. It is primarily based entirely on reactive criticism design, which may be a type of reactive criticism design (A.C.D.). The goal is to keep the application's functionality to a minimum while adhering to strict constraints. The parameters of both networks were retrained using the particle swarm optimization (PSO) method [35–37]. This blended approach will significantly increase planning accuracy and range compared to previous methods using A.D.C.s without PSO.

Smart home gadgets can be used in new ways given the fact that they can travel with power from the client to the network and vice versa. Under certain circumstances, a smart meter can determine the power usage for a character's household equipment. These facts can be collected, stored and evaluated (HEMS) [38]. HEMS is a tool (i.e., primarily based on time of use pricing) that works from home and allows you to reserve equipment to buy money and effort. However, family equipment cannot be programmed, so we cannot control all of it. Refrigerators, lights, and electric powered kettles are examples of non-schedulable gadgets that cannot be scheduled because of their function. Appliances that may be programmed (which include a bathing system, a dryer, and a home warm water boiler) may be in addition divided into:

- *Interruptible*: Items that may be turned off at any time (i.e., a home warm water boiler, an electric-powered car charger);
- *Non-interruptible*: Items that cannot be turned off once turned on (which include a bathing system or dishwasher).

These examples display how smart grids rely upon smart home to function. Smart appliances, collectively with statistics from diverse sensors and smart meters, permit a smart domestic home to optimize its quantitative and temporal strength intake from the grid, slicing strength costs. The scope of one of these surveys prevents an in-depth evaluation of this thing of smart home. Curious readers can locate statistics from numerous assets in literature [39–41].

2.3   RELATED SOFTWARE AND SMART HOME

Following an overview of key features of smart home and device operating systems describing two primary software configurations: data processing and occupant behavior tracking.

2.3.1   OPERATING SYSTEM

Internet of Things (IoT) technology is connected by various communication methods and integrated with a microcontroller unit (MCU)-based system. The software does play an essential part in IoT functioning. End devices and gateways are the two categorizations of IoT elements that are integrated with the operating system.

Terminals, including sensors, actuators, and switches, are examples of the most effective shutdown devices to perform a limited number of tasks. These devices are often small, have resource-constrained MCUs, are very environmentally friendly and have relatively short communication times. The MCU contains built-in tools pre-installed in ROM. The tools are not changed or updated at the factory. In contrast, MCUs are particularly cheap to deploy while at the same time becoming more complex and performing better (e.g., migrating from 8/16 to 32-bit architecture). As a result, you might have a software program that can perform additional work on the tool itself and receive real-time protection updates (OTAs). This software program is probably a fully functional system. This will improve the safety and reliability of the tool.

Because the information must be acquired and transmitted in authentic time with no buffering, such an operating system is referred to as a real-time operating system. And besides, the available resources to such end systems are restricted (RTOS). Because the OS serves as a gateway to many reduced processes, using RTOS helps a developer or systems designer be more efficient [42]. More effective machines (portals) are essential (drivers) to accumulate information from those stop structures (detectors) and to ship indicators that function on the idea of events.

Gateway gadgets help more than one conversation protocol and are fantastically able to accept and study information, as a consequence appearing as a hyperlink among many IoT structures in domestic automation. When cloud offerings constitute the idea of smart domestic design, those gateway gadgets are also referred to as part gateways because they're placed at the interface among the outside the Internet and the inner personal intranet. This makes those gadgets and OS more effective. They must also be safeguarded and resistant to attacks from the outside world via the Internet.

Unlike end systems, portal units typically provide an interface that allows the user to operate the networks or see data. The following are the most significant elements to contemplate when choosing an OS for the Internet of Things [43–45]:

- *Footprint*: Due to the restricted resources of most end devices, the OS must have a small footprint to function on. This issue may not be relevant for gateway devices, which often have many resources.
- *Scalability*: The OS must separate the available resources or CPU architecture.
- *Modularity*: The working gadget should be adaptable so that gadget analysts can select which additives to consist of. The working gadget commonly gives the handiest the minimum capability needed to push a utility to the MCU. If extra packages are required (which includes a guide for different verbal exchange protocols), the working gadget ought to include them as standard.
- *Portability*: The OS must run on various hardware platforms, perhaps with different architectures. Typically, this is achieved by abstracting hardware requirements.
- **Connectivity:** An IoT device's capacity to interact with other gadgets is critical. The OS must support Ethernet, Wi-Fi, Bluetooth LE, and other communication technologies.
- *Security*: The OS must allow for the addition of security features as needed. Lighter systems come with fewer security measures by default, but the developer must add more if necessary.
- *Reliability*: Because most end systems are not frequently maintained, the OS must function without errors for extended periods. This factor implies they must also satisfy specific certifications, particularly if they are utilized in mission-critical circumstances.

The capacity to execute several applications simultaneously is another key feature of an OS. This feature is chiefly essential for R.T.O.S. because operations must meet all requirements and should not spend exorbitant time queuing for assets to be allocated. Various programming models define how the program is executed and how processes are assigned assets. Different algorithms exist depending on the reason for which the OS was created.

2.3.2 TRACKING OF HUMAN ACTIVITY

Intelligent housing needs to strengthen the attitude of residents towards their lives or optimize their functions. In any situation, the ability of the device to evaluate and record the behavior of the device population is very important [13]. Video cameras, infrared cameras, radar systems, floor tension sensors, and various media for calling hobbies are just a few examples. Activity IDs need to appear in real time so that home automation can quickly react and adapt as needed. This phenomenon can be achieved by fog and facet computing, which we will investigate later. Motion detection and prediction is a two-step process. The first is the discovery of hobbies made with undefeated technology. The second phase uses supervised learning techniques to recognize and predict activities, aided by the clustered information collected in the former step.

2.3.2.1 Methods of Following and Detecting Unobtrusive

Sensing and monitoring must be non-intrusive for intelligent homes to provide a better degree of convenience and ease. To put it another way, the inhabitants should have no idea they're being watched or compelled to wear particular body sensors at all times. Numerous important advancements have been made in this approach utilizing wireless signals. This technique was initially created for military uses, such as detecting individuals behind barriers or tracking them down [46–47].

On the other hand, the technology focused on military-grade relied on high-power wireless transmissions and allocated wireless spectrum, making it unsuitable for general usage. MIT researchers have devised a novel method for measuring respiration and heart rate that use a radar technology known as FMCW (frequency-modulated carrier waves). The method may identify breathing-related chest motions and heartbeat-related skin vibrations [48]. Thanks to strategically placed sensing devices, which can trace the individual at a distance up to 8 meters away, surprisingly at the back of a wall, this is enough shield for a typical home. The accuracy is even more impressive: 99.3% for respiration and 98.5% for heartbeat. Doppler radars have been used in similar experiments. Li et al. [49–50] conducted a detailed review of advancements in radar sensors [51]. Accelerometric sensors placed on the mattress are another option.

While inconspicuous vital signal detection techniques are mainly being developed for medical applications, their full potential can be realized in other domains. A smart home's purpose is to utilize "intelligence" to create welcoming and flexible surroundings. Home automation systems with more precise information about occupants (e.g., heart and breathing rates) can respond faster and provide more relaxation. For example, the ideal temperature in a home is presently estimated without considering the people who live there by analyzing interior and exterior temps, humidity levels, tension and other environmental variables. If their heart rate and respiration rates were considered while determining the ideal temperature in residence, better decisions might be reached. These vital markers are linked to surface temperatures and general health [52].

2.3.2.2 Activity Discovery

Activity discovery is an essential aspect of human activity tracking. Because the different sensory devices (cameras, motion detectors) will continually create data streams, effective algorithms that can recognize new activities and subsequently be utilized for recognition must be built. Unsupervised learning methods, to our understanding, are top suited in identifying patterns in huge volumes of data which also concludes to the summary of activity discovery. Cook et al. [13] employ an unsupervised study method that divides the input stream into smaller groups, making the activity detection process easier. The data for detecting activity is compressed and stored as a series of information from several sensors.

The Kpattern clustering approach turned into evolved via a method by Bourobou et al. [53], who discovered pastime styles for substantial quantities of facts gathered via detectors. The K sample contrasts with different distinguished paradigms and Kmeans, Maximization, Expectation and Fastest First. They discovered that Kmuster turned to different merchandise in phrases of execution time and the number of clusters detected.

2.3.2.3   Activity Detection and Prediction

Activity detection and prediction offer sensible surroundings the capacity to behave on what the person is doing. Suppose a person enters a cubicle after 10 pm, sits on a mattress and remains for a while. In such situations, the smart home can dim the lighting, assuming that the person has fallen asleep while the temperature has dropped to a comfortable sleeping level. Scholars are presently operating on mixing numerous techniques and frameworks for equipping smart home with pastime identity and prediction capabilities.

Cook et al. [13] examined Support Vector Machine (SVM), Hidden Markov Model (HMM), I Bayes Classifier, and Conditional Random Field. SVM became out to be an excellent performer. In addition to enhancing performance, it again becomes a shape level for a selected occasion type. An available method cannot offer a general solution, so specific strategies must be incorporated to obtain higher effects and extra state-of-the-art hybrid smart systems [54]. For example, genetic algorithms (GAs) or primarily herd-based technology are nicely applicable to enhance the system studying and predictive abilities of devices [55].

Recognition of human pastime via way of means of Jalal et al. [56] must be run the usage of an intensity camera (Microsoft Kinect). Scientists selected the HMM for studying because the identity of sports is primarily based totally on a chain of actions. They used the movement coding representations that they had gathered earlier. With an accuracy of 92.33%, the authors located that an HMM with four hidden states gave great effects.

Fatima et al. used SVM-based kernel classifiers as well as conditional random fields [57]. It presents a common framework that permits domestic automation to stumble on and manipulate pastimes. The accuracy of pastime detection changed from 92.70% to 94.11%, at the same time as the overall performance of pastime prediction via way of means of the writer changed from 79.71% to 84.78%. The developing wide variety of IoT generates a sizable quantity of data, necessitating the improvement of current foundations and technology to manipulate and save it.

2.3.3   Data Collection and Usage

Three variables describe large records: the "5 Vs" [58–59] of volume, velocity, variety, value and veracity.

- *Volume* is the number of records created per second from various resets that require storage and analysis.
- The rate at which that data is generated, gathered, and analyzed is called the *velocity*.
- Data *heterogeneity* is described by different types of records (based and unstructured) referred as *variety*.
- Data *fidelity* is the degree to which a record is flawlessly reliable, affecting its use and relevance, referred as *veracity*.
- The term *cost* refers to the usefulness or *value*. This is related to accuracy as the cost of records increases and the desire to ensure integrity and accuracy increases.

2.3.3.1 Data Collection

In-home automation records can be obtained from one of three possible options: active interaction with the customer, passive communication with the consumer, or raw data not generated by the consumer. Data generated directly through the client is a living source for interacting with the consumer, including voicing commands, gesture tracking, and an action by pressing a key or indulging with a prompt interface [60–64].

2.3.3.2 Fusion of Data

Data gathered by numerous sensing devices in an intelligent home must be combined and analyzed to regain important information about the environment and the condition of the occupants. You can use data fusion technology that "combines data from multiple sensors with relevant information from a linked database to provide more accurate and detailed results than with a single sensor alone" [65]. The technique of combining data from multiple gadgets is called multi-sensor fusion. This is an inherently tricky process due to the ever-increasing variety of data and the amount of hands-on knowledge.

The average weight is the most straightforward statistical technique, although multivariate statistical analysis and the most cutting-edge data mining technologies are also available [66]. When combined with estimators or classifiers with varying shows for incomputable data, the statistical method may not be suitable [67]. Maximum likelihood techniques [68–69], filtering by Kalman and [70], as well as the theory of evidence, are examples of probabilistic methodologies. The Kalman filter is often used with low complexity, easy implementation, and optimal mean square error.

Unfortunately, it cannot be used with data where it is difficult to characterize the defect attributes. Artificial intelligence technologies include genetic algorithms, neural networks, and decision trees. In many applications [66–67], neural networks are used as a classifier construction and data fusion approach. There are two approaches to data fusion: control and distribution. In this scenario, the data from the source is merged according to a set of criteria at the central point of central fusion (such as a smart home gateway). In distributed fusion, data is fused at the source (the sensor) and requires processing power [71].

```
                    ┌─────────────────────┐
                    │   Communication     │
                    │    protocols        │
                    └─────────────────────┘
          ┌──────────────────┼──────────────────┐
┌──────────────────┐ ┌──────────────────┐ ┌──────────────────┐
│     Wired        │ │    Wireless      │ │  Combination of  │
│  communication   │ │  communication   │ │   Wireless and   │
│    protocols     │ │    protocols     │ │  Wired Protocols │
└──────────────────┘ └──────────────────┘ └──────────────────┘
```

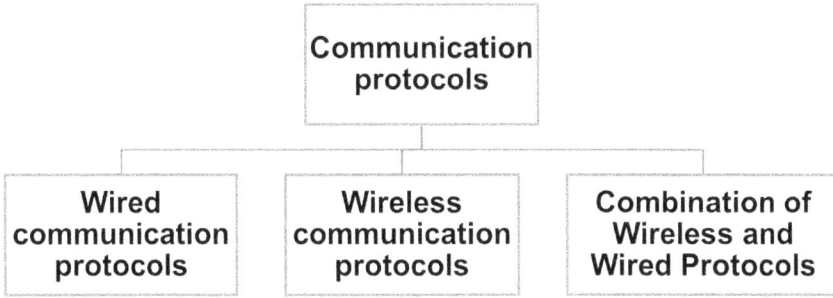

FIGURE 2.2   Category of communication protocols.

2.4   CONNECTION AND COMMUNICATION PROTOCOL

Devices should be linked to the speaking records in a smart home. Intelligence takes place while the surroundings are familiar with the system's modern nation, as stated above. In this scenario, one sensor on its own is insufficient to extract many beneficial records. This way, we want a massive range of sensors to speak with every different devices. The mechanism with the aid of using which numerous gadgets and sensors can speak is described using the conversation protocol. Organizations and coalitions explain standards, hardware requirements, software requirements, and licenses for these protocols, which control how data is delivered [72–73]. Communication protocols are split into three groups based on transmission as follows (Figure 2.2).

The application situation determines the optimum technology to use. Some protocols have a more extended range, better security, and less power than others. Furthermore, the scale of the network has a role in the selection.

2.4.1   WIRED COMMUNICATION PROTOCOLS

Wired communication points to the transfer of data through corded means. It dates back to the days when texts were being sent via electrical telegraph, and it is one of the earliest forms of communication. Table 2.1 shows the wired communication protocols.

TABLE 2.1
Wired Communication Protocols

	Ethernet	X10	UPB	INSTEON	MoCA	KNX
Frequency	100–500 MHz	120 kHz; 310–433.92 MHz	4–40 kHz	131.62; 868–924 MHz	0.5–1.5 GHz	110/132 kHz; 863.3 MHz
Data Rate	1 MBpx-100 Gbps	20–6- bps; 9.6 kbps	490 bps	13.165 kbps; 38.4 kbps	175 Mbps-2.5 Gbps	1.2/2.4 Mbps
Range	100 m	500–1000m	80–500 m	500 m; 40m	90 m	1000 m; 100m
Network Topology	Bus, Star	None, Star	P2P	P2P, Mesh, dual Mesh	P2P, Mesh	Tree, Line Star
Encryption	None	None	None	AES-256	DES-56, AES-128	None, AES -128

- *Ethernet* is one of the most popular technologies in wired LAN and WAN networks [74], based on the IEEE 802.3 standard. It ranges up to 100 meters and is resistant to electromagnetic interference.
- *X10* is extensively diagnosed because of the first frequent conversation protocol for sign transduction and management of patron electronics. Data is transmitted through the usage of strong lines. X10 has many top-notch drawbacks when utilized in smart modern homes (e.g., low facts rates, low cable connectivity and absence of encryption support, most range of linked gadgets, interference and messages gap) [75–78].
- *UPB*: UPB (Universal Powerline Bus) is a patented era that communicates via strength lines, much like X10. Specifics include quicker facts throughput (even though nonetheless slower than different technologies), decreased noise from AC strength lines, and an elevated range of gadgets activated with the aid of using the student community (as much as 64,000) in comparison to the X10. Lack of encryption and reputation in constrained markets also are disadvantageous [79].
- *INSTEON*: It is a unique hybrid era that can control smart home devices from a distance using both wires and RF transmission. Because it uses a mesh architecture, no critical hubs are required, and all INSTEON devices can optionally communicate with each other and repeat messages to ensure a growing community. The advantages are reliability, ease of use, interoperability, fast messaging, and various gadgets. On the other hand, epochs have a gradual rate of language exchange, which is helpful in controlling gadgets but not now useful for moving records [75] [79].
- MoCA (multimedia over coax): It is a method to ensure multimedia distribution in the home using coaxial cable. It is a robust and reliable oral communication protocol with a packet error rate of only 106% [75]. It also strengthens Wi-Fi insurance while maintaining record performance in terms of Wi-Fi repeaters. MoCA 2.5 has a theoretical maximum write rate of 2–5 Gbps.
- KNX: It is a standardized and fully OSI-based oral communication protocol designed for the smart home (EN 50090, ISO/IEC 14543). KNX is best suited for signal and instrument control because of its complexity and cost, requiring its wiring and low registration fees. It supports three supported topologies: linear, tree and star. It is a community that brings people together. Multiple transmission media are supported.

The following are some of the *benefits* of wired data transfer versus wireless transmission:

- *Security*: Because the entire network necessitates connecting the gadget with a cord, eavesdropping or tampering with the network information from the outside is extremely difficult.
- *User-friendliness*: Linking to a network is as simple as inserting the cord into the gadget; unlike wireless networks, there is no want to pick a community from a listing or input a passcode.

- *Distance*: Data transmitted throughout a cord will move longer than not unusually placed Wi-Fi protocols; because cords are an encapsulated medium, interference and impediments will not impact the transmission.
- *Data rate*: Ethernet has a theoretical maximum throughput of 100 Gbps, while Wi-Fi 802.11ac has a theoretical maximum throughput of 1.3 Gbps.
- *Reliability*: Data transfer by cable is reliable and unaffected by interference or blockages; nevertheless, transmission rates on wireless networks may fluctuate.

On the other hand, wired communication techniques have several *drawbacks*:

- *Complexity and cost*: Building a wired network requires specialized work and planning. Integrating wired networks into smart homes must be done when building a home. The subsequent laying of cables on the wall can be cumbersome and uncomfortable.
- *Mobility*: It is impossible to relocate the gadget once the wires have been attached without expanding the cable or cord.
- *Electricity*: Normally, wired connections require electricity to function; however, if power is lost in a critical situation, the network will not function. It's possible that the network won't be as battery friendly as a cordless network.
- *Expansions*: Expanding the range of network wiring requires more effort than just installing a new wireless router and may necessitate the purchase of extra hardware.

2.4.2 WIRELESS COMMUNICATION PROTOCOLS

The use of radio frequency waves to send and receive signals without wires is known as wireless communication. Because of their easy usage and inexpensive charges of putting in a community and including new gadgets, Wi-Fi communications technology has become increasingly popular in smart domestic networks. Table 2.2 shows the wireless communication protocols.

TABLE 2.2
Wireless Communication Protocols

	Wi-Fi 802.11n	Bluetooth	Bluetooth LE	ZigBee	Z-Wave	6LowPAN
Frequency	2.4–5.8 GHz	2.402–2.48 GHz	2.402–2.48 GHz	868/915 MHz, 2.4 GHz	868/915 MHz	868/921 MHz, 2.4–5Ghz
Data Rate	450 Mbps	0.7–2.1 Mbps	2 Mbps	20–40 kbps, 250 kbps	10–100 kbps	10–40 kbps, 250 kbps
Range	10–100 m	15–20 m	10–15 m	10–100 m	30–50 m	10–100 m
Network Size	Thousand (Mesh)	8	N/A	65,536	232	250
Network Topology	Star, Tree, P2P, Mesh	Star	Star	Star, Tree, Mesh, Cluster Tree	Mesh	Star, Mesh, P2P
Encryption	WPA2	AES-128	AES-128	AES-128	AES-128	AES-128

- *Wi-Fi*: Wireless fidelity or Wi-Fi is described using the IEEE 801.11 standard. It is one of the most popular Wi-Fi technologies as it no longer requires a license. The theoretical coverage is 45 m indoors, but the use of Wi-Fi repeaters and redundant hotspots can extend the range, making it suitable for WLANs. This era supports WPA2 encryption and operates in the 2.4- to 5.8-GHz frequency range. There are downsides to Wi-Fi, including that it is vulnerable to excessive power consumption and interference. In addition, the overall performance and reliability of a community may be affected by internal barriers.
- *Bluetooth*: Bluetooth is a broadly used Wi-Fi technology for personal area networks. Bluetooth operates on frequencies of 2402–2480 MHz and 2400–3483 MHz. There are 79 channels, each with 1 MHz bandwidth. However, a few nations observe channel restrictions. Thanks to the unfolding spectrum (FHSS), this is sufficient to keep away from channel congestion and channel switching [80]. Bluetooth is a usually used Wi-Fi method for structures and gadgets. Bluetooth Low Energy (BLE) is a subset of Bluetooth designed for low-electricity gadgets that may run on battery strength for lengthy durations of time. Bluetooth five introduces some enhancements to the BLE version, inclusive of broader range, higher channel selection, and better records rates, with a focal point on enhancing IoT device support,
- *ZigBee*: ZigBee is a low-rate wireless communication system based on the IEEE 802.15.4 standard that is designed for equipment with limited power supplies. Since this technique is open-source and freely available to everyone, it is a favorite option among lower power device makers. The data rate is typically 20–250 kbps with a 70-meter impact area [74] [79]. It permits increasing the range of activity, using multi-hop propagation; however, it can often cause the "popcorn effect," which is detained inactivity caused by the text transmitted through a gadget to another prior to reaching the last recipient.
- *Z-Wave*: Z-Wave is another low-power technology that has been designed to be dependable. Unlike ZigBee, Z-Wave is a proprietary technology that requires a license and certification from the Z-Wave Alliance [80]. With a range of up to 50 meters, transfer rates of up to 100 kbps are feasible. Z-Wave may be used to create mesh networks, which means that devices can communicate without the need for a central gateway or controller. If a device is not nearby, notifications can hop up to four times across nodes to reach their destination [81].
- **6LowPAN:** 6LowPAN is an available specification for building low-power IPv6 PANs and an open source. Data rates of 20–250 kbps, ranging from 10 to 100 m, are supported depending on the frequency. Because IPv6 is used, each device has its IPv6 address and may be reached through the Internet. It is a 50-company collaboration to establish the 6LowPAN technique as part of smart home inter-process communication [82].

Compared to wired communication, wireless communication offers various advantages. Because a device does not require a physical connection to connect to a network, it may be moved about without losing connections and straightforward switching to a different cordless network. A few *advantages* of wireless communications protocols include:

- *Scalability*: It's easy to add gadgets to a community if you do not exceed the number of gadgets supported. The Wi-Fi network is clean, and you are free to scale up or down for free or at a minimal cost.
- *Cost*: Organizing a Wi-Fi community is straightforward and can be completed regularly without the help of an expert.
- *Flexibility*: Organizing your Wi-Fi community with new features is as easy as connecting to your device, so you can properly test it in different gadgets and sensor locations.

The following are some of the *disadvantages* of wireless communication:

- *Safety*: Despite modern encryption mechanisms, packets travel over the air and can be intercepted and decoded (although unlikely). Most security issues arise when the Wi-Fi community is not covered in any way due to the loss of enough settings.
- Wi-Fi networks are theoretically slower than under-stressed networks (including Ethernet or MoCA), but actual data rates are usually good enough for the smartest home applications.
- *Intervention*: Wireless networks are susceptible to interference that can cause service disruption or degraded performance.
- *Coverage*: Wireless networks theoretically provide more coverage in a given area than stress-prone networks. However, due to the perimeter or improper placement of the instrument, insurance may be limited, and commands/messages may be lost3.

2.5   PRIVACY AND SECURITY

Without omnipresent processors and a swarm of detectors spread throughout a house, the notion of a smart home system would not be viable. Nevertheless, the usage of such gadgets, which are commonly hooked up to the Internet (directly or indirectly) and employ wireless technology, opens new opportunities for cyberattacks on connected home inhabitants' confidentiality [83–84].

Data privacy and security: Security in terms of home automation is primarily concerned with data security and privacy and the occupants' privacy. Privacy and physical security are important. Security must address the following concerns, according to Zheng et al. [85]:

1. Data breach prevention (ensuring that unauthorized parties are unable to access data)
2. Authorization (the process of deciding who has entry to data
3. Keeping user's privacy safe.

The most often used approach for data protection is strict and asymmetric key cryptography. Substitute procedures use biometric parameters such as nerve inter-pulse intervals and vascular blood volume. Because these measurements may be challenging for small portable equipment, these approaches have energy efficiency and processing power limits. Internal and external dangers are the most common risks to a smart home. Internal assaults are feasible when the cybercriminal is close to the residence. Exterior assaults are feasible when the cybercriminal has access to the Internet. In either case, the attacker wants to get access to information saved on cloud services or damage the smart home's architecture. A cybercriminal might employ several typical risks against an intelligent home and its occupants [86–89], including the following:

Eavesdropping: If an assailant gains entry to a router, then one can monitor the traffic in and out of the building, jeopardizing the residents' anonymity and safety. If the assailant comes near enough to the house, they could use specialized equipment to eavesdrop messages sent by a variety of sensing devices and gadgets, and they could use specialized equipment to decrypt texts sent from a variety of sensing devices and devices.

Impersonation: This type of danger arises when a hacker tries to make choices as a spokesperson for a client either by stealing qualifications and property or by executing a man-in-the-middle plan. Because of eavesdropping technology, such situations are probable. The lawbreaker obtains entry to the suspect's information and then manipulates or repeats network appeals to carry out suspicious assaults.

Software exploitation: Many consumers will put up a gadget as a smart home architecture but will fail to take the necessary steps to change the default administration password. Another danger in this class is neglecting to keep software up to date and secure; this permits hackers to enter holes in the gadget and gain administrator entry, which may lead to more attack opportunities.

Denial of service (DoS): Intruders hinder the normal working of sensing devices or routers by making repeated appeals to the device at a single time or transmitting garbled texts that the device cannot handle, causing the device to fail. Hackers can disable Internet connectivity in a smart home in this fashion, preventing owners from using the Internet to access their home.

Ransomware is a relatively new type of cyberattack. Hackers get into a computer, encrypt the data on the discs with an encryption key, and then demand payment in return for the encryption key so they may decode the data.

2.6 FUTURE TRENDS AND CHALLENGES

Every home automation device stakeholder must be fully engaged to eliminate the hazard hurdles. In a larger IOT context, the ENISA (European Union Agency for Network and Information Security) has created a compendium highlighting the dangers and potential solutions to magnify the present state of cybersecurity in intelligent home surroundings. The entities are essential for keeping a smart home system safe and secure from external rebukes.

For IoT-based smart dwellings, Risteska Stojkoska and Trivodaliev [3] highlight many problems, issues and their answers. In the realm of edge (fog) computing, the authors emphasize the necessity for speeding interaction among Smart Home gadgets and advocating the construction of ultra-light pseudo-codes for local information processing. The vast volumes of data collected by the sensing devices require a decrease in the number of gadgets exchanging new big data technologies for aggregation, archiving and evaluation.

The difficulty of interoperability that comes with connectivity is now being addressed through the development of rules to guarantee that different providers supply handsets. The last point to consider regarding the compatibility of IoT-based smart homes is privacy and safety. The slow adoption of smart homes by customers is a significant barrier in and of itself [90]. To define the rate of acceptance of smart homes, Shin et al. evolved a technological adoption replica. The essential variables in making a purchase decision, according to their research, are compatibility, enabling circumstances of usage, and perceived utility.

People also argue that elderly clients are more minded to acquire intelligent homes than younger ones. According to the survey, in order to grow vendor requests, a master plan geared at younger clients is needed. Consumer acceptability is intimately linked to how smart homes are used and how they deliver on pleasure, simplicity, safety and relaxation, as well as power generation, are all promises [91]. In a qualitative study, Hargreaves et al. investigated the accessibility of home automation gadgets. The authors emphasize four key aspects of intelligent home technology:

1. Applied and collective disruption;
2. Household adaption and familiarization;
3. Educating to use is difficult, and there is little assistance;
4. There is no large power reduction potential and there are perils of power escalation.

The researchers go on to examine the study's larger practical, scientific and theoretical implications, stating that smart home adoption must go beyond new technology, taking into consideration individual user biographies to reflect the broader influence on their everyday lives and activities [92–93]. In the future, the smart home will connect a huge number of gadgets to the grid, allowing them to take part in the system- and local-level coordinating activities. These gadgets' computing capacity and data might be employed to continue the active supply and demand stability under the transitive energy (TE) paradigm [94]. The pliability of abundant producing and load assets in TE systems is used to achieve this equilibrium [95]. Using real-time, decentralized decision-making, TE may benefit the whole grid system while admiring the interests and behaviors of individual members.

Dorri et al. [96] proposed lightweight blockchain implementation geared to IoT-based smart home. According to the authors, increased security and privacy may be accomplished with minimum traffic, processing time or energy consumption. Sensors, appliances and other equipment allow smart home technology. On the other hand, smart house management solutions determine smart home performance and service quality. To adapt to varied smart home usage scenarios and user expectations, an adaptable development framework is required. As proposed by Xu et al., software-defined smart dwellings provide flexibility and ease of deployment [97].

Sensors, appliances and other equipment allow smart home technology. Smart home management systems, on the other hand, evaluate the performance and quality of smart homes. To adapt to varied smart home usage scenarios and user expectations, an adaptable development framework is required. As proposed by Xu et al. [97], software-defined smart dwellings provide this flexibility while also being simple to build. The proposed framework is based on virtualization, transparency and centralized design ideas, allowing for effective integration of diverse smart home devices and compatibility and standards. This adaptable framework's unique components are location-based home automation, changeable healthier lifestyles and smart home condition observation.

2.7 CONCLUSIONS

Home automation is no longer a far-fetched fantasy. There has been a significant amount of study and investment, and various smart homes have been examined and installed throughout the world. However, due to cost, complexity and a lack of expertise, buying a smart house is still not a popular choice. As a result, accessibility surveys [91] and targeted information techniques, in addition to continuous research and development operations, are crucial to expanding customer requirements [90]. Simple yet effective layouts are crucial for the widespread adoption of smart homes. Device compatibility, smart home network set-up and its complexity, and the non-appearance of standardized associates for the management of devices are all problems that must be resolved. The multiplicity of accessible devices perplexes consumers; as a result, big vendors must agree on technical stacks. Ordinary homeowners are uninterested in technical aspects that may improve their homes' comfort, enjoyment and power efficiency [98]. Some recent advancements, such as application-oriented smart buildings, may help with adaptability and compatibility difficulties [97].

Finally, home automation is the result of many people working together. Engineers experiment with and design portals for activities like embedded sensors, deep learning for behavior disclosure, identification, predictions for the future and methods to engage with a multitude of data arriving from sensing devices, as well as techniques to merge the data to derive usable information. Intelligent home initiatives, on the contrary, will necessitate the participation of additional experts in order for them to be well-integrated systems that can adapt to tenant demands and external occurrences and change with their occupants' evolving preferences [96] [99].

Builders, construction researchers, bankers, sociologists, and other professionals must be involved in designing future smart homes that will be embraced by consumers and break through the currently limited market. Smart grids, smart suburbs, intelligent buildings, smart administrations and, eventually, a smart planet will all be possible thanks to them.

REFERENCES

[1]. M.R. Alam, M.B.I. Reaz, M.A.M. Ali, A review of smart homes-past, present, and future, IEEE Trans. Syst. Man Cybern. Part C (Appl. Rev.) 42 (6) (2012), 1190–1203, doi:10.1109/TSMCC.2012.2189204.

[2]. S.J. Darby, Smart technology in the home: time for more clarity, Build. Res. Inf. 46 (1) (2018), 140–147, doi:10.1080/09613218.2017.1301707.

[3]. B.L. Risteska Stojkoska, K.V. Trivodaliev, A review of internet of things for the smart home: Challenges and solutions, J. Clean. Prod. 140 (2017), 1454–1464, doi:10.1016/J. JCLEPRO.2016.10.006.

[4]. T. Kramp, R. van Kranenburg, S. Lange, Introduction to the internet of things, in Enabling Things to Talk, Springer, Berlin, Heidelberg, 2013, pp. 1–10, doi:10.1007/978-3-642-40403-0_1

[5]. Y. Strengers, Smart Energy Technologies in Everyday Life: Smart Utopia? Springer, Berlin, 2013.

[6]. W.K. Edwards, R.E. Grinter. At home with ubiquitous computing: Seven challenges. in G.D. Abowd, B. Brumitt, S. Shafer (eds), Ubicomp 2001: Ubiquitous Computing. Ubi-Comp 2001. Lecture Notes in Computer Science, vol 2201, Springer, Berlin, Heidelberg, doi:10.1007/3-540-45427-6_22.

[7]. D.J. Cook, S.K. Das, Smart Environments: Technologies, Protocols, and Applications, John Wiley, Hoboken, 2005.

[8]. S. Das, D. Cook, A. Battacharya, E. Heiserman, Tze-Yun Lin, The role of prediction algorithms in the MavHome smart home architecture, IEEE Wirel. Commun. 9 (6) (2002), 77–84, doi:10.1109/MWC.2002.1160085.

[9]. R. Blasco, Á. Marco, R. Casas, D. Cirujano, R. Picking, A smart kitchen for ambient assisted living, Sensors 14 (12) (2014), 1629–1653, doi:10.3390/s140101629.

[10]. F. Buttussi, L. Chittaro, MOPET: A context-aware and user-adaptive wearable system for fitness training, Artif. Intell. Med. 42 (2) (2008), 153–163, doi:10.1016/J. ARTMED.2007.11.004.

[11]. A. Jalal, J.T. Kim, T.S. Kim, Development of a Life Logging System via Depth Imaging-based Human Activity Recognition for Smart Homes, 2012, http://citeseerx.ist. psu.edu/viewdoc/download?doi=10.1.1.456.9125&rep=rep1&type=pdf

[12]. M. Soliman, T. Abiodun, T. Hamouda, J. Zhou, C.H. Lung, Smart home: Integrating Internet of things with web services and cloud computing, in Proceedings of the IEEE Fifth International Conference on Cloud Computing Technology and Science, IEEE, Piscataway, 2013, pp. 317–320, doi:10.1109/CloudCom. 2013.155.

[13]. D.J. Cook, A.S. Crandall, B.L. Thomas, N.C. Krishnan, CASAS: A smart home in a box, Computer 46 (7) (2013), doi:10.1109/MC.2012.328.

[14]. Y. Jie, J.Y. Pei, L. Jun, G. Yun, X. Wei, Smart home system based on IoT technologies, in Proceedings of the International Conference on Computational and Information Sciences, IEEE, Piscataway, 2013, pp. 1789–1791, doi:10.1109/ICCIS.2013.468.

[15]. J. Zhou, T. Leppanen, E. Harjula, M. Ylianttila, T. Ojala, C. Yu, H. Jin, CloudThings: A common architecture for integrating the Internet of Things with Cloud Computing, in Proceedings of the IEEE Seventeenth International Conference on Computer Supported Cooperative Work in Design (CSCWD), IEEE, Piscataway, 2013, pp. 651–657, doi:10.1109/CSCWD.2013.6581037.

[16]. J. Hosek, P. Masek, D. Kovac, M. Ries, F. Kröpfl, IP home gateway as a universal multi-purpose enabler for smart home services. Elektrotech. Informationstechnik 131 (4–5) (2014), 123–128.

[17]. S. Guoqiang, C. Yanming, Z. Chao, Z. Yanxu, Design and Implementation of a Smart IoT Gateway, in Proceedings of the IEEE International Conference on Green Computing and Communications and IEEE Internet of Things and IEEE Cyber, Physical and Social Computing, IEEE, 2013, pp. 720–723, doi:10.1109/GreenCom-iThings-CPSCom.2013.130.

[18]. P.M. Mell, T. Grance, SP 800–145. The NIST Definition of Cloud Computing, Technical Report, National Institute of Standards and Technology, Gaithersburg, MD, US, 2011, https://csrc.nist.gov/publications/detail/sp/800-145/final

[19]. L.D. Xu, W. He, S. Li, Internet of things in industries: A survey, IEEE Trans. Ind. Inf. 10 (4) (2014), 2233–2243, doi:10.1109/TII.2014.2300753.

[20]. T. Goyat, N. Pandey, V.K. Shukla and A.V. Singh, Review of protocol security and cloud model in cloud computing environment, in 2021 9th International Conference on Reliability, Infocom Technologies and Optimization (Trends and Future Directions) (ICRITO), 2021, pp. 1–6, doi:10.1109/ICRITO51393.2021.9596402.

[21]. Fog Computing and the Internet of Things: Extend the Cloud to Where Things Are, www.cisco.com/c/dam/en_us/solutions/trends/iot/docs/computing-overview.pdf

[22]. M. Chiang, T. Zhang, Fog and IoT: An overview of research opportunities, IEEE Internet Things J. 3 (6) (2016), 854–864, doi:10.1109/JIOT.2016.2584538.

[23]. M. Aazam, E.N. Huh, Fog computing and smart gateway based communication for cloud of things, in Proceedings of the International Conference on Future Internet of Things and Cloud, IEEE, Piscataway, 2014, pp. 464–470, doi:10.1109/FiCloud.2014.83.

[24]. S. Yi, Z. Hao, Z. Qin, Q. Li, Fog Computing: platform and applications, in Proceedings of the Third IEEE Workshop on Hot Topics in Web Systems and Technologies (Hot-Web), IEEE, Piscataway, 2015, pp. 73–78, doi:10.1109/HotWeb.2015.22.

[25]. S. Yi, Z. Qin, Q. Li, in Security and Privacy Issues of Fog Computing: A Survey, Springer, Cham, 2015, pp. 685–695, doi:10.1007/978-3-319-21837-3_67.

[26]. M. Ganzha, L. Maciaszek, M. Paprzycki, Proceedings of the 2016 federated conference on computer science and information systems: September 11–14, 2016. Gdańsk, Poland, in Federated Conference on Computer Science and Information Systems (No. 8). Institute of Electrical and Electronics Engineers, Polskie Towarzystwo Informatyczne, 2016.

[27]. C. Carvalho, The Gap between Processor and Memory Speeds, https://pdfs.semantic-scholar.org/6ebe/c8701893a6770eb0e19a0d4a732852c86256.pdf

[28]. Fog vs. Edge Computing: What's the Difference? http://info.opto22.com/fog-vs-edgecomputing

[29]. P. Garcia Lopez, A. Montresor, D. Epema, A. Datta, T. Higashino, A. Iamnitchi, M. Barcellos, P. Felber, E. Riviere, Edge-centric computing, ACM SIGCOMM Comput. Commun. Rev. 45 (5) (2015), 37–42, doi:10.1145/2831347.2831354.

[30]. W. Shi, J. Cao, Q. Zhang, Y. Li, L. Xu, Edge Computing: vision and challenges, IEEE Internet Things J. 3 (5) (2016), 637–646, doi:10.1109/JIOT.2016.2579198.

[31]. W. Shi, S. Dustdar, The promise of edge computing, Computer 49 (5) (2016), 78–81, doi:10.1109/MC.2016.145.

[32]. U.S. Congress, Energy Independence and Security Act of 2007, December 18, 2007. https://www.epa.gov/laws-regulations/summary-energy-independence-and-security-act [Online], Accessed on May 2023.

[33]. X.H. Li, S.H. Hong, User-expected price-based demand response algorithm for a home-to-grid system, Energy 64 (2014), 437–449, doi:10.1016/J.ENERGY.2013.11.049.

[34]. D. Fuselli, F. De Angelis, M. Boaro, S. Squartini, Q. Wei, D. Liu, F. Piazza, Action dependent heuristic dynamic programming for home energy resource scheduling, Int. J. Electr. Power Energy Syst. 48 (2013), 148–160, doi:10.1016/J.IJEPES.2012.11.023.

[35]. T. Huang, D. Liu, Residential energy system control and management using adaptive dynamic programming, in Proceedings of the International Joint Conference on Neural Networks, IEEE, Piscataway, 2011, pp. 119–124, doi:10.1109/IJCNN.2011.6033209.

[36]. R.L. Welch, G.K. Venayagamoorthy, Optimal control of a photovoltaic solar energy system with adaptive critics, in Proceedings of the International Joint Conference on Neural Networks, IEEE, Piscataway, 2007, pp. 985–990, doi:10.1109/IJCNN.2007.4371092.

[37]. R. Welch, G. Venayagamoorthy, Comparison of two optimal control strategies for a grid-independent photovoltaic system, in Proceedings of the Conference Record of the 2006 IEEE Industry Applications Conference Forty-First IAS Annual Meeting, 3, IEEE, Piscataway, 2006, pp. 1120–1127, doi:10.1109/IAS. 2006.256673.

[38]. B. Zhou, W. Li, K.W. Chan, Y. Cao, Y. Kuang, X. Liu, X. Wang, Smart home energy management systems: Concept, configurations, and scheduling strategies, Renew. Sustain. Energy Rev. 61 (2016), 30–40, doi:10.1016/J.RSER.2016.03.047.

[39]. P. Siano, Demand response and smart grids – A survey, Renew. Sustain. Energy Rev. 30 (2014), 461–478.

[40]. M.L. Tuballa, M.L. Abundo, A review of the development of smart grid technologies, Renew. Sustain. Energy Rev. 59 (2016), 710–725.

[41]. H.T. Haider, O.H. See, W. Elmenreich, A review of residential demand response of smart grid, Renew. Sustain. Energy Rev. 59 (2016), 166–178.

[42]. W. Lamie, The Benefits of RTOSes in the Embedded IoT | EE Times, www.eetimes.com/author.asp?section_id=36&doc_id=1327623

[43]. A. Milinković, S. Milinković, L. Lazic, Choosing the right RTOS for IoT platform, Infoteh-Jahorina 14 (2015), 504–509.

[44]. IoT Operating Systems, https://devopedia.org/iot-operating-systems

[45]. T. Reusing, Comparison of operating systems Tinyos and Contiki, Sens. Nodes-Operation, Netw. Appli. (SN) 7 (2012).

[46]. R. Zetik, S. Crabbe, J. Krajnak, P. Peyerl, J. Sachs, R. Thomä, in Detection and Localization of Persons Behind Obstacles Using M-sequence Through-the-Wall Radar, 6201, International Society for Optics and Photonics, 2006, p. 62010I, doi:10.1117/12.667989.

[47]. A.R. Hunt, In: A wideband imaging radar for through-the-wall surveillance, 5403, International Society for Optics and Photonics, 2004, 590, doi:10.1117/12.542718.

[48]. F. Adib, H. Mao, Z. Kabelac, D. Katabi, R.C. Miller, Smart homes that monitor breathing and heart rate, in Proceedings of the Thirty-Third Annual ACM Conference on Human Factors in Computing Systems – CHI '15, ACM Press, New York, NY, USA, 2015, pp. 837–846, doi:10.1145/2702123.2702200.

[49]. O. Postolache, P.S. Girão, R.N. Madeira, G. Postolache, Microwave FMCW doppler radar implementation for in-house pervasive health care system, in IEEE International Workshop on Medical Measurements and Applications, IEEE, Piscataway, 2010, pp. 47–52.

[50]. C. Li, V.M. Lubecke, O. Boric-Lubecke, J. Lin, A review on recent advances in doppler radar sensors for noncontact healthcare monitoring, IEEE Trans. Microwave Theory Tech. 61 (5) (2013), 2046–2060, doi:10.1109/TMTT.2013.2256924.

[51]. F. Studnicka, P. Seba, D. Jezbera, J. Kriz, Continuous monitoring of heart rate using accelerometric sensors, in Proceedings of the Thirty-Fifth International Conference on Telecommunications and Signal Processing, TSP 2012, 2012, pp. 559–561.

[52]. P. Davies, I. Maconochie, The relationship between body temperature, heart rate and respiratory rate in children, Emer. Med. J. 26 (9) (2009), 641–643, doi:10.1136/emj.2008.061598.

[53]. S. Bourobou, Y. Yoo, User activity recognition in smart homes using pattern clustering applied to temporal ANN algorithm, Sensors 15 (5) (2015), 11953–11971, doi:10.3390/s150511953.

[54]. B. Qela, H.T. Mouftah, Observe, Learn, and Adapt (OLA) an algorithm for energy management in smart homes using wireless sensors and artificial intelligence, IEEE Trans. Smart Grid 3 (4) (2012), 2262–2272, doi:10.1109/TSG.2012.2209130.

[55]. P. Rocca, M. Benedetti, M. Donelli, D. Franceschini, A. Massa, Evolutionary optimization as applied to inverse scattering problems, Inverse Probl. 25 (12) (2009), 123003, doi:10.1088/0266-5611/25/12/123003.

[56]. A. Jalal, S. Kamal, D. Kim, A depth video sensor-based life-logging human activity recognition system for elderly care in smart indoor environments, Sensors 14 (12) (2014), 11735–11759, doi:10.3390/s140711735.

[57]. I. Fatima, M. Fahim, Y.K. Lee, S. Lee, A unified framework for activity recognition-based behavior analysis and action prediction in smart homes, Sensors 13 (2) (2013), 2682–2699, doi:10.3390/s130202682.

[58]. E.H. Pflugfelder, Big data, big questions, Communication Design Quarterly Review 1(4) 2013, 18–21.

[59]. S. Sagiroglu, D. Sinanc, Big data: A review, in 2013 International Conference on Collaboration Technologies and Systems (CTS), IEEE, Piscataway, 2013, pp. 42–47.

[60]. A. Yazan, W. Yong, N. Raj Kumar, Big data life cycle: Threats and security model, in 21st Americas Conference on Information Systems, Association for Information Systems (AIS), Puerto Rico, 2015.

[61]. S. K. Dubey, S. Mittal, S. Chattani, V. K. Shukla, Comparative analysis of market basket analysis through data mining techniques, in 2021 International Conference on Computational Intelligence and Knowledge Economy (ICCIKE), IEEE, Dubai, UAE, 2021, pp. 239–243, doi:10.1109/ICCIKE51210.2021.9410737.

[62]. R. Punhani, V.P.S. Arora, A.S. Sabitha, V.K. Shukla, Segmenting e-commerce customer through data mining techniques, in Journal of Physics: Conference Series, Volume 1714, 2nd International Conference on Smart and Intelligent Learning for Information Optimization (CONSILIO), Goa, India, 24–25 October 2020.

[63]. T.F. Kappukalar Nasurudeen, V.K. Shukla, S. Gupta, Automation of disaster recovery and security in cloud computing, 2021 International Conference on Communication information and Computing Technology (ICCICT), IEEE, Mumbai, India, 2021, pp. 1–6, doi:10.1109/ICCICT50803.2021.9510110.

[64]. A. Gandomi, M. Haider, Beyond the hype: Big data concepts, methods, and analytics, International Journal of Information Management 35 (2) (2015), 137–144.

[65]. D. Hall, J. Llinas, An introduction to multisensor data fusion, Proc. IEEE 85 (1) (1997), 6–23, doi:10.1109/5.554205.

[66]. J. Han, M. Kamber, Data Mining: Concepts and Techniques, Elsevier, Amsterdam, 2012.

[67]. S. Hashem, Sherif, optimal linear combinations of neural networks, Neural Netw. 10 (4) (1997), 599–614, doi:10.1016/S0893-6080(96)00098-6.

[68]. D. Huang, H. Leung, An expectation maximization-based interacting multiple model approach for cooperative driving systems, IEEE Trans. Intell. Transp. Syst. 6 (2) (2005), 206–228, doi:10.1109/TITS.2005.848366.

[69]. A. Mohammad-Djafari, Probabilistic methods for data fusion, in Maximum Entropy and Bayesian Methods, Springer, Dordrecht, Netherlands, 1998, pp. 57–69, doi:10.1007/978-94-011-5028-6_5.

[70]. D. Dubois, H. Prade, Possibility Theory, Springer, Boston, MA, US, 1988, doi:10.1007/978-1-4684-5287-7.

[71]. M. Mitici, J. Goseling, M. de Graaf, R.J. Boucherie, Decentralized vs. centralized scheduling in wireless sensor networks for data fusion, in Proceedings of the IEEE International Conference on Acoustics, Speech and Signal Processing (ICASSP), IEEE, Piscataway, 2014, pp. 5070–5074, doi:10.1109/ICASSP.2014.6854568.

[72]. R. Ghai, S. Singh, An architecture and communication protocol for picocellular networks, IEEE Personal Communications Magazine 1(3) (1994), 36–46.

[73]. M. Popovic, Communication Protocol Engineering, CRC Press, Boca Raton, 2018.

[74]. M. Kuzlu, M. Pipattanasomporn, S. Rahman, Review of communication technologies for smart homes/building applications, in Proceedings of the IEEE Innovative Smart Grid Technologies – Asia (ISGT ASIA), IEEE, Piscataway, 2015, pp. 1–6, doi:10.1109/ISGT-Asia.2015.7437036.

[75]. T. Mendes, R. Godina, E. Rodrigues, J. Matias, J. Catalão, Smart home communication technologies and applications: Wireless protocol assessment for home area network resources, Energies 8 (7) (2015), 7279–7311, doi:10.3390/en8077279.

[76]. M. Li, H.J. Lin, Design and implementation of smart home control systems based on wireless sensor networks and power line communications, IEEE Trans. Ind. Electr. 62 (7) (2015), 4430–4442, doi:10.1109/TIE.2014.2379586.

[77]. O. Bello, S. Zeadally, Network layer inter-operation of Device-to-Device communication technologies in Internet of Things (IoT), Ad Hoc Netw. 57 (2017), 52–62, doi:10.1016/J. ADHOC.2016.06.010.

[78]. P. Darbee, Insteon Whitepaper: Compared, http://cache.insteon.com/documentation/ insteon_compared.pdf

[79]. M. Poulakis, S. Vassaki, G. Pitsiladis, C. Kourogiorgas, A. Panagopoulos, G. Gardikis, S. Costicoglou, Wireless sensor network management using satellite communication technologies, in Emerging Communication Technologies Based on Wireless Sensor Networks, CRC Press, Boca Raton, 2016, pp. 201–232, doi:10.1201/b20085–12.

[80]. O. Horyachyy, Comparison of Wireless Communication Technologies Used in a Smart Home: Analysis of Wireless Sensor Node Based on Arduino in Home Automation Scenario, 2017, www.diva-portal.org/smash/get/diva2:1118965/FULLTEXT02

[81]. C. Withanage, R. Ashok, C. Yuen, K. Otto, A comparison of the popular home automation technologies, in Proceedings of the IEEE Innovative Smart Grid Technologies – Asia (ISGT ASIA), IEEE, Piscataway, 2014, pp. 600–605, doi:10.1109/ ISGT-Asia.2014.6873860.

[82]. Z. Shelby, C. Bormann, 6LoWPAN: The Wireless Embedded Internet, John Wiley & Sons, Hoboken, 2011.

[83]. S.F.N. Zaidi, V.K. Shukla, V.P. Mishra, B. Singh, Redefining home automation through voice recognition system. In A.E. Hassanien, S. Bhattacharyya, S. Chakrabati, A. Bhattacharya, S. Dutta (eds), Emerging Technologies in Data Mining and Information Security. Advances in Intelligent Systems and Computing (vol 1300), Springer, Singapore, 2021, doi:10.1007/978-981-33-4367-2_16.

[84]. S. Ibrahim, V. K. Shukla, R. Bathla, Security enhancement in smart home management through multimodal biometric and passcode, in 2020 International Conference on Intelligent Engineering and Management (ICIEM), Institute of Electrical and Electronics Engineers (IEEE), London, United Kingdom, 2020, pp. 420–424, doi:10.1109/ ICIEM48762.2020.9160331.

[85]. Y.L. Zheng, X.R. Ding, C.C.Y. Poon, B.P.L. Lo, H. Zhang, X.L. Zhou, G.Z. Yang, N. Zhao, Y.T. Zhang, Unobtrusive sensing and wearable devices for health informatics, IEEE Trans. Biomed. Eng. 61 (5) (2014), 1538–1554, doi:10.1109/ TBME.2014.2309951.

[86]. D. Geneiatakis, I. Kounelis, R. Neisse, I. Nai-Fovino, G. Steri, G. Baldini, Security and privacy issues for an IoT based smart home, in Proceedings of the Fortieth International Convention on Information and Communication Technology, Electronics and Microelectronics (MIPRO), IEEE, Piscataway, 2017, pp. 1292–1297, doi:10.23919/ MIPRO.2017.7973622.

[87]. C. Lee, L. Zappaterra, Kwanghee Choi, Hyeong-Ah Choi, Securing smart home: Technologies, security challenges, and security requirements, in Proceedings of the IEEE Conference on Communications and Network Security, IEEE, Piscataway, 2014, pp. 67–72, doi:10.1109/CNS.2014.6997467.

[88]. F.F. Petiwala, I. Mearaj, V.K. Shukla, Analyzing cyber security breaches, in ICT and Data Sciences, CRC Press, Boca Raton, 2022, pp. 63–72.

[89]. S. Gupta, M. Kumar, S. Bhushan, V.K. Shukla, Risk analysis assessment of interdependency of vulnerabilities: in cyber-physical systems, in *Holistic Approach to Quantum Cryptography in Cyber Security*, CRC Press, Boca Raton, pp. 227–234.

[90]. J. Shin, Y. Park, D. Lee, Who will be smart home users? An analysis of adoption and diffusion of smart homes, Technol. Forecast. Soc. Change 134 (2018), 246–253.

[91]. T. Hargreaves, C. Wilson, R. Hauxwell-Baldwin, Learning to live in a smart home, Build. Res. Inf. 46 (1) (2018), 127–139.

[92]. V. K. Shukla, B. Singh, Conceptual framework of smart device for smart home management based on RFID and IoT, in 2019 Amity International Conference on Artificial Intelligence (AICAI), IEEE, Dubai, UAE, 2019, pp. 787–791, doi:10.1109/AICAI.2019.8701301.

[93]. R. Khan, V.K. Shukla, B. Singh, S. Vyas, Mitigating security challenges in smart home management through smart lock, in T.P. Singh, R. Tomar, T. Choudhury, T. Perumal, H.F. Mahdi (eds), Data Driven Approach Towards Disruptive Technologies. Studies in Autonomic, Data-driven and Industrial Computing, Springer, Singapore, 2021, doi:10.1007/978-981-15-9873-9_7.

[94]. S. Chen, C. Liu, From demand response to transactive energy: State of the art, J. Mod. Power Syst. Clean Energy 5 (1) (2017), 10–19.

[95]. M. Marzband, F. Azarinejadian, M. Savaghebi, E. Pouresmaeil, J.M. Guerrero, G. Lightbody, Smart transactive energy framework in grid-connected multiple home microgrids under independent and coalition operations, Renew. Energy 126 (2018), 95–106.

[96]. A. Dorri, S.S. Kanhere, R. Jurdak, P. Gauravaram, Blockchain for IoT security and privacy: The case study of a smart home, in Proceedings of the IEEE International Conference on Pervasive Computing and Communications Workshops, PerCom Workshops 2017, 2017, pp. 618–623.

[97]. K. Xu, X. Wang, W. Wei, H. Song, B. Mao, Toward software defined smart home, IEEE Commun. Mag. 54 (5) (2016), 116–122.

[98]. C. Links, What is SHaaS? And why should you care? Qorvo White Paper, 2016, www.zigbee.org/wp-content/uploads/2016/11/Qorvo-Whitepaper-ShaaS.pdf

[99]. K. Christidis, M. Devetsikiotis, Blockchains and smart contracts for the internet of things, IEEE Access 4 (2016), 2292–2303.

3 Home Automation for Urban Infrastructure

*Komal Saxena, Yagya Buttan
and Vinod Kumar Shukla*

CONTENTS

3.1 INTRODUCTION

In this fast-paced world, we want a few smaller things to be done at our fingertips to make life easier without putting small effort into those little things. That is why the concept of home automation is getting very popular [1]. We can control our home applications using our phones or voice control to save time and energy and make life flexible and easier. To the most extent, home automation is successful and promising. However, at the same time, it is very challenging to use for those who are unaware or not comfortable with using these technologies.

Referring towards the physically disabled population, old age population, and dependent on others for their day-to-day work in their places, where they need a system that they can understand and use easily, Figure 3.1 presents the old age population in various countries [2]. This demonstrates a need to address these age groups related to home automation. In old age, problems like heart complaints, nervous disorders, asthma, tuberculosis and other problems are very common among a large percentage of the population. This requires efficient home automation-related technologies, which this population can use easily [3]. Disability is something anyone of us can go through, irrespective of age. It can be a temporary or permanent disability. There can be various disabilities like disabilities due to hearing, vision, physical, cerebral, learning and so on. Hearing disability can be such that a person may have a lower hearing capacity or sometimes permanent hearing disability like being deaf and mute. Vision disability makes people difficult to read such that they use spectacles. Some may not be able to see anything, making them blind. Physical disability may arise due to loss of hands or legs by birth or car accident [4].

DOI: 10.1201/9781003218715-4

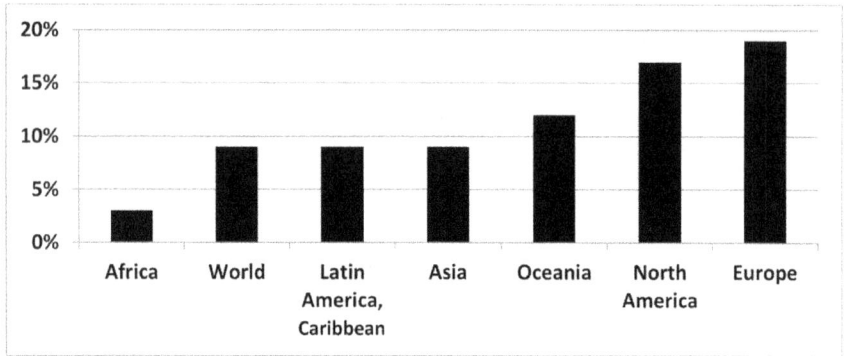

FIGURE 3.1 Proportion of old age people as of world population.

The home automation system can be designed in such a way that people with determination can be benefitted. And would be able to control basic electrical appliances such as lights, fans, television, air conditioners, and many other related services/product with just a normal microcontroller such as Arduino [5]. The home automation market is increasing rapidly, and by the next few years will serve many people; therefore, it will not only look for people who can use a smartphone or control things using voice.

This chapter presents the concept of a device that can help older and disabled people or people who are, due to some reason, dependent on other people to control their house electrical appliances using easy-to-remember gestures. The architecture of this device consists of a small microcontroller Arduino Nano with Bluetooth [6–7] for receiving signals and a remote with a gesture recognizer sensor with an Arduino Nano and Bluetooth for sending what the user wants to do.

3.2 GESTURE CONTROL AND HOME AUTOMATION

Home automation with the help of IoT devices has helped to reduce energy consumption via smart monitoring of the use of devices like air-conditioners which are climate controlled based on the room temperature, thus functioning efficiently. Similarly, light systems used with IoT technology help in learning the need for the amount of brightness in rooms such that during the day, the light is not in full brightness, but when it is dark or at night, the light brightens to its total capacity [8]. The older people who have issues walking and moving from places like a bed in order to switch off lights might be an issue, but with the help of a smart automation home system, they can switch on and off or even control the functioning of lights, fans and air-conditioners [8].

Home automation [9] is a system of devices in homes that are smart and interconnected with other devices and can be controlled by an application or a smart device like smartphones and tablets. Gesture control helps better control these devices as there is no need for one to switch on and off devices manually, but with a hand gesture, they can increase or decrease the speed of fans and brightness of lights.

Gesture controls have limitations, especially when hand gestures are used. This is due to the several image cues like color, texture and shape used to approach the problem of hand detection. To solve the issue of lighting conditions, active color models are maintained using a neural network-based face detection system [10]. Gestures are one of the technologies that have helped us easily use our daily devices like mobile phones in the context of pervasive computing. These include speech recognition, gesture recognition and so on. Gesture technology also allows us to use our custom, personalized gestures. For example, we can set certain gestures to open a specific application on our mobile phone using gestures [11–12].

3.3   LITERATURE REVIEW

Numerous household-based automation systems have been newly generated in the current scenario. Few of them have been generally transformed to assist with the requirements of the visually impaired and old age individuals. They use a sensors microcontroller which can be Arduino/Raspberry Pi computer hardware, and the central hub, which is generally operated by the mobile app. Some of the specific characteristics fetch affluence of existing and also caring for individuals who cannot fight for themselves and cope with these new systems. The enactment of home-based automation can be imprecisely separated into different Methods. It can be intelligent, voice-based, likely to be available market goods or founded on WSN (Wireless Sensor Networks). The projection of the gesture-controlled systems [13] is interrelated to the home automation systems of the voice-controlled system, which uses the old traditional off-and-on devices from verbal commands. In [14], the author has alienated the system into different dual fragments: wireless device and smart voice recognition. It is based on an Android App which generates voice instructions from the operator, and a proper process is made. The disadvantages of intelligent-based systems for the old and visually impaired are that they are unable to speak clearly. It also seems that these people are not very well versed with smartphones and henceforth might not be able to completely use the system made for their support in the most effective way. In [15], the authors proposed a design comprising a light controlling system, a notification alerting through a vibrating system and detection the gas system using sensors. Many existing systems help achieve the ultimate goal of automating house systems. In [16], the author is focused on the Remote Intelligent houses that have adequate size to detect and monitor the household devices which can check humidity or temperature.

Similarly, many of the proposed systems employ wireless sensor networks (WSNs), hosted on private Wi-Fi networks using ZigBee In [17–18]. These networks may consist of different types of sensors like temperature sensors [19], pressure sensors on beds [20], water usage monitoring sensors [21] or even sound sensors. Optimization of these networks to minimize sensors and power is discussed in [22]. Studies have also been done to compare efficiency and feasibility [23]. In this article, the author discusses that using smart houses is valuable. These systems remotely control sensor which is being used to detect movement. In these types of intelligent homes, the control is not in the hand of the user. In case the system fails or is not accurate, it could generate a wastage of energy and show turbulent behavior. It decrypts

ASL (American Sign Language) signs using ANN (artificial neural networks) which will be operated and controlled by domestic appliances and devices. The raw data is pre-processed and decoded. Zigbee is to be used for transmission. Monitoring activities of the elderly in their own homes remotely through a mobile application is done in [24]. In [25], glove-based gesture-controlled, a combination of accelerometer and flex sensors improved the efficiency.

Home automation is not a new term today, but indeed is growing and booming in the industry with its new application in the real world. This technology is very useful in smart cities [26]. In this, we are monitoring health through IoT, electronic-based devices which we can use to alert the system and send reminders [27]. However, most of them work either with the use of the Internet or smart devices such as a smartphone which does not cover people who are disabled, old peoples who cannot understand how to use new technology or people such as with Aphasia disorder. The voice-enabled automation system is available, but even those cannot cover people who cannot speak or have pronunciation problems like lispers or those with rhotacism. There are other conditions where people cannot use such automation.

Sensors and IoT technologies are playing very important in real-time data capturing and real-time data analysis such as supply chain [28–29], transport and passenger management [30–32], tourism [33], agriculture management [34–36], healthcare [37–38], and many more domains.

The proposed framework gives a simple, easy-to-follow solution for such people. These systems require a good amount of programming, whereas this system is easy to build and cost-efficient.

3.4 FRAMEWORK USING GESTURE SENSORS FOR HOME AUTOMATION

The system includes a special PAJ7320 gesture sensor [39] made for gesture recognition, including nine different types of basic gestures that are easy to remember. These different gestures can be used to control appliances according to the requirements given by the user at the time of installation. For example, if a user rotates their finger clockwise once on the top of the sensor fan will start working, and doing the same second time will stop the fan. Basic gestures include backward, clockwise, wave, up, down, left, right, forward and counterclockwise. There are multiple possibilities for new gestures according to need.

3.5 ARCHITECTURE FOR DEVICE

The architecture of the device includes two different sides: the receiver side and the sender side.

The receiver side (Figure 3.2A) includes an open-source microcontroller Arduino Nano having a constant power supply with a relay connected to various appliances such as fans, lights, television and many more, as well as a Bluetooth receiver. The sender side (Figure 3.2B) includes a remote having a gesture sensor connected to Arduino Nano and Bluetooth for sending signals whenever the user performs any gesture. If the user performs any gesture, the sender side will send a signal to the

BLUETOOTH RECEIVER

APPLIANCES   RELAY   Arduino Nano

Power Supply

FIGURE 3.2A Block diagram—receiver side.

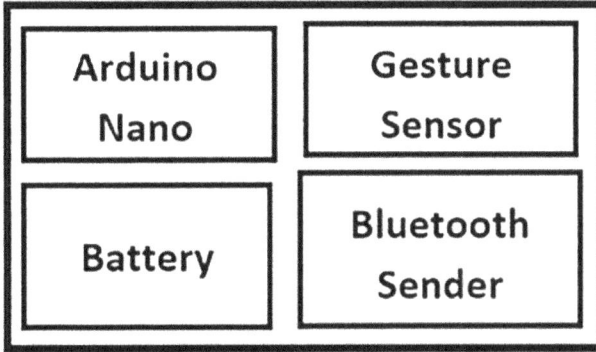

Arduino Nano | Gesture Sensor

Battery | Bluetooth Sender

FIGURE 3.2B Block diagram—sender side.

receiver side that has to be performed, and the appliance will start or stop working according to the need [40]. The system uses a specially made gesture sensor made for human hand gestures and development with microcontroller, which makes it easy to handle and portable, so it can be used anywhere in a defined range.

3.5.1 HARDWARE REQUIREMENTS

1. *Arduino*—An open-source microcontroller [34]. A platform for building an electronics project using the physical programmable board and an integrated development environment (IDE) with large community support. Arduino board have various types of boards such as Arduino Mega, Uno, Nano, Lilypad and many more. The board used in the system is Arduino Nano based on an Atmega328P chipset with 14 digital pins, 8 analog input pins, 16 MHz clock speed, 5 V operating voltage and 1 KB EEPROM (Figure 3.3).
2. *PAJ7620 gesture sensor*—General I2C interface into one single chip for gesture recognition function. It can identify basic nine gestures including in all the possible movements of move down, move left, up, move, etc. with a simple swipe of the hand [41], and we can create more possible gestures (Figure 3.4).
3. *Relay*—Switch electrically operable, contains an input terminal for the signal.
4. *Bluetooth hc05 or hc06*—Module for Bluetooth connectivity (Figure 3.5).

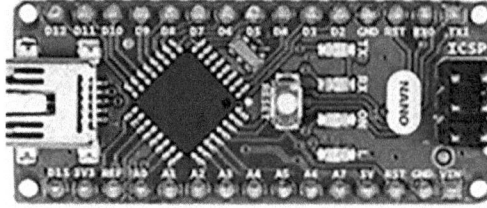

FIGURE 3.3   Arduino Nano–based board. The board used in the system is Arduino Nano based on Atmega328P chipset with 14 digital pins, 8 analog input pins, 16 MHz clock speed, 5 V operating voltage and 1 KB EEPROM.

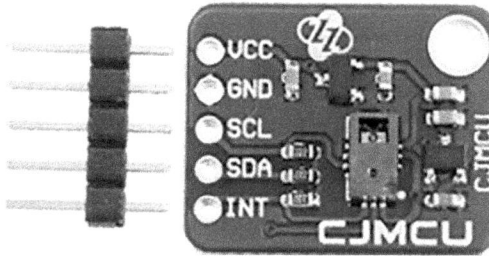

FIGURE 3.4   PAJ7620 gesture sensor.

FIGURE 3.5   Bluetooth hc06/hc05.

3.6   FUTURE SCOPE AND CONCLUSION

The proposed model and system are made for people dependent on others, for example, people in wheelchairs, older people, people who cannot move due to medical conditions such as those on complete bed rest, and people who cannot speak. A major transformation is underway in the senior care market. The rising demand for eldercare services is the largest trend it is facing. Big data and the technology of artificial intelligence will play an increasingly significant part in the future of senior care in the near future. The future scope of this proposed framework can also use artificial intelligence and machine learning methods to change the products design to train products for the home automation system to make more advances.

REFERENCE

[1]. Khade, N.S., et al., 2017. "Vandana publications home automation." *International Journal of Engineering and Management Research (IJEMR)* 7.1: 390–392.

[2]. Statista, *Proportion of Selected Age Groups of World Population in 2020, by Region* [Online]. www.statista.com/statistics/265759/world-population-by-age-and-region/ [Accessed on: 19–05–2021]

[3]. Balamurugan, J., and Ramathirtham, G., 2012. Health problems of aged people. *International Journal of Research in Social Sciences*, 2.3: 139.

[4]. Narayanan, S., 2018, October. A study on challenges faced by disabled people at workplace in Malaysia. In *Proceeding—5th Putrajaya International Conference on Children, Women, Elderly and People with Disabilities* (pp. 185–197), International Journal For Studies On Children, Women, Elderly And Disabled People, Malaysia.

[5]. Malav, V., et al., 2019. "Research paper on Bluetooth based home automation using Arduino." *Journal of Advancements in Robotics* 6.2: 9–14.

[6]. Kavitha, J., et al., 2018. "Bluetooth based home automation using Arduino and Android application." *International Journal for Research in Applied Science and Engineering Technology* 6.3: 2003–2009.

[7]. David, N., Chima, A., Ugochukwuand, A., and Obinna, E., 2015. "Design of a home automation system using Arduino." International journal of Scientific & Engineering Research 6: 795–801.

[8]. Vishwakarma, S.K., Upadhyaya, P., Kumari, B., and Mishra, A.K., 2019, April. Smart energy efficient home automation system using IoT. In *2019 4th International Conference on Internet of Things: Smart Innovation and Usages (IoT-SIU)* (pp. 1–4). IEEE.

[9]. Arathi, P.N., Arthika, S., Ponmithra, S., Srinivasan, K. and Rukkumani, V., 2017. Gesture based home automation system. In *2017 International Conference on Nextgen Electronic Technologies: Silicon to Software (ICNETS2)* (pp. 198–201). IEEE.

[10]. Hackenberg, G., McCall, R., and Broll, W., 2011, March. Lightweight palm and finger tracking for real-time 3D gesture control. In *2011 IEEE Virtual Reality Conference* (pp. 19–26). IEEE

[11]. Liu, J., Zhong, L., Wickramasuriya, J., and Vasudevan, V., 2009. uWave: Accelerometer-based personalized gesture recognition and its applications. *Pervasive and Mobile Computing* 5.6: 657–675.

[12]. Poppinga, B., Sahami Shirazi, A., Henze, N., Heuten, W., and Boll, S., 2014, September. Understanding shortcut gestures on mobile touch devices. In *Proceedings of the 16th International Conference on Human-Computer Interaction with Mobile Devices & Services* (pp. 173–182). Association for Computing Machinery (ACM) Digital Library.

[13]. Mittal, Y., Paridhi, T., Sonal, S., Deepika, S.L., Ruchi, G., and Vinay Kumar, M., 2015. "A voice-controlled multifunctional smart home automation system." In *2015 Annual IEEE India Conference (INDICON)* (pp. 1–6). IEEE.

[14]. Ravi, A., Brindha, R., Janani, S., Meenatchi, S., and Prathiba, V., 2018. "Smart voice recognition based home automation system for aging and disabled people." *International Journal of Advanced Scientific Research & Development (IJASRD)* 5.1: 11–18.

[15]. Ansah, A.K., Jeffrey Antwi, A., and Stephen, A., 2015. "Technology for the aging society-a focus and design of a cost-effective smart home for the aged and disabled." *Proceedings of the World Congress on Engineering and Computer Science* 1.

[16]. Bindroo, O., Saxena, K., and Khatri, S.K., 2017, August. A wearable NFC wristband for remote home automation system. In *2017 2nd International Conference on Telecommunication and Networks (TEL-NET)* (pp. 1–6). IEEE.

[17]. Hossain, M.S., Md Abdur, R., and Ghulam, M., 2017. "Cyber–physical cloud-oriented multi-sensory smart home framework for elderly people: An energy efficiency perspective." *Journal of Parallel and Distributed Computing* 103: 11–21.

[18]. Ransing, R.S., and Manita, R., 2015. "Smart home for elderly care, based on Wireless Sensor Network." In *2015 International Conference on Nascent Technologies in the Engineering Field (ICNTE)* (pp. 1–5). IEEE.

[19]. Suryadevara, N.K., Mukhopadhyay, S.C., Ruili, W., and Rayudu, R.K., 2013. "Forecasting the behavior of an elderly using wireless sensors data in a smart home." *Engineering Applications of Artificial Intelligence* 26.10: 2641–2652.

[20]. Gaddam, A., Mukhopadhyay, S.C., and Sen Gupta, S.C., 2010. "Smart home for elderly using optimized number of wireless sensors." In *Advances in Wireless Sensors and Sensor Networks* (pp. 307–328). Springer.

[21]. Gaddam, A., Mukhopadhyay, S.C., and Sen Gupta, S.C., 2009. "Smart home for elderly care using optimized number of wireless sensors." In *2009 4th International Conference on Computers and Devices for Communication (CODEC)* (pp. 1–4). IEEE.

[22]. Chernbumroong, S., Atkins, A., and Yu, H., 2010. "Perception of smart home technologies to assist elderly people." In *4th International Conference on Software, Knowledge, Information Management and Applications* (pp. 90–97). Paro, Bhutan. ISBN: 978-974-672-556-9

[23]. Prabhuraj, R., and Saravanakumar, B., 2014. "Gesture controlled home automation for differently challenged people." *International Journal of Research in Electronics* 1.2: 1–6.

[24]. Basit, A., Saxena, K., and Rana, A., 2020. "A wearable device used for smart doorbell in home automation system." In *2020 IEEE International Women in Engineering (WIE) Conference on Electrical and Computer Engineering (WIECON-ECE)* (pp. 90–93). Bhubaneswar, India, doi:10.1109/WIECON-ECE52138.2020.9398001

[25]. Fahim, M., Iram, F., Sungyoung, L., and Young-Koo, L., 2012. "Daily life activity tracking application for smart homes using Android smartphone." In *2012 14th International Conference on Advanced Communication Technology (ICACT)* (pp. 241–245). IEEE.

[26]. Moirangthem, P., Saxena, K., Basit, A., and Rana, A., 2020. "Explorative state-wise study of smart cities in India." In *2020 8th International Conference on Reliability, Infocom Technologies and Optimization (Trends and Future Directions) (ICRITO)* (pp. 1–5). Noida, India, doi:10.1109/ICRITO48877.2020.9197789.

[27]. El-Basioni, Mohammad, B.M, Abd El-Kader, S.M., and Eissa, H.S., 2014. "Independent living for persons with disabilities and elderly people using smart home technology." *International Journal of Application or Innovation in Engineering and Management* 3.4: 11–28.

[28]. Wanganoo, L., and Shukla, V.K., 2020, July. Real-time data monitoring in cold supply chain through NB-IoT. In *2020 11th International Conference on Computing, Communication and Networking Technologies (ICCCNT)* (pp. 1–6). IEEE.

[29]. Wanganoo, L., Shukla, V.K., and Panda, B.P., 2021. "NB-IoT powered last-mile delivery framework for cold supply chain." In Singh T.P., Tomar R., Choudhury T., Perumal T., and Mahdi, H.F. (eds), *Data Driven Approach Towards Disruptive Technologies. Studies in Autonomic, Data-driven and Industrial Computing*. Springer, Singapore, doi:10.1007/978-981-15-9873-9_22

[30]. Madana, A.L., and Shukla, V.K., 2020, June. Conformity of accident detection using drones and vibration sensor. In *2020 8th International Conference on Reliability, Infocom Technologies and Optimization (Trends and Future Directions)(ICRITO)* (pp. 192–197). IEEE.

[31]. Siraj, A., and Shukla, V.K., 2020. "Framework for personalized car parking system using proximity sensor." In *2020 8th International Conference on Reliability, Infocom Technologies and Optimization (Trends and Future Directions) (ICRITO)* (pp. 198–202). doi:10.1109/ICRITO48877.2020.9197853.

[32]. Madana, A.L., Shukla, V.K., Sharma, R., and Nanda, I., 2021, March. IoT enabled smart boarding pass for passenger tracking through bluetooth low energy. In *2021 International Conference on Advance Computing and Innovative Technologies in Engineering (ICACITE)* (pp. 101–106). IEEE.

[33]. Verma, A., Shukla, V.K., and Sharma, R., 2021. Convergence of IOT in tourism industry: A pragmatic analysis. In *Journal of Physics: Conference Series* (Vol. 1714, No. 1, p. 012037). IOP Publishing.

[34]. Shukla, V.K., Kohli, A., and Shaikh, F.A., 2020. "IOT based growth monitoring on moringa oleifera through capacitive soil moisture sensor." In *2020 Seventh International Conference on Information Technology Trends (ITT)* (pp. 94–98). doi:10.1109/ITT51279.2020.9320884.

[35]. Nanda, I., Sahithi, C., Swath, M., Maloji, S., and Shukla, V.K., 2020. "IIOT based smart crop protection and irrigation system." In *2020 Seventh International Conference on Information Technology Trends (ITT)* (pp. 118–125). doi:10.1109/ITT51279.2020.9320783.

[36]. Murlidharan, S., Shukla, V.K., and Chaubey, A., 2021. "Application of machine learning in precision agriculture using IoT." In *2021 2nd International Conference on Intelligent Engineering and Management (ICIEM)* (pp. 34–39). doi:10.1109/ICIEM51511.2021.9445312.

[37]. Trayush, T., Bathla, R., Saini, S., and Shukla, V.K., 2021. "IoT in healthcare: Challenges, benefits, applications, and opportunities." In *2021 International Conference on Advance Computing and Innovative Technologies in Engineering (ICACITE)* (pp. 107–111). doi:10.1109/ICACITE51222.2021.9404583.

[38]. Shukla, V.K., and Verma, A., 2019. "Enhancing user navigation experience, object identification and surface depth detection for "low vision" with proposed electronic cane." In *2019 Advances in Science and Engineering Technology International Conferences (ASET)* (pp. 1–5). doi:10.1109/ICASET.2019.8714213.

[39]. *Seeed, Grove—Gesture V1.0* [Online]. https://wiki.seeedstudio.com/Grove-Gesture_v1.0/ [Accessed On: 08–06–2021]

[40]. *Martyn Currey, Arduino to Arduino by Bluetooth* [Online]. www.martyncurrey.com/arduino-to-arduino-by-bluetooth/ [Accessed On: 08–06–2021]

[41]. *Hackster, Hand Gesture Recognition Sensor (PAJ7620)* [Online]. hackster.io/SurtrTech/hand-gesture-recognition-sensor-paj7620–9be62f [Accessed On: 08–06–2021]

4 Optimizing LPWAN Using Particle Swarm for the Smart City

Sindhu Hak Gupta, Asmita Singh, Abhishek Tyagi and Jitendra Singh Jadon

CONTENTS

4.1 INTRODUCTION

A smart city uses smart, efficient and sustainable technologies [1]. Over time, technology progressively paved the way for development and has played an important role in enabling the growth of nations. Fog and edge computing are rapidly advancing with cloud computing [2]. Until recently, cloud-level processing adopted a centralized mechanism, which inadvertently resulted in high network traffic and delays while processing bulk data. As mentioned, a more robust approach has been leveraged late to overcome the drawback, involving a decentralized approach of data processing at the network's nodes (edge computing) or distributed levels (fog computing), which shot up the cloud storage space. With the advent of complex smart city models, the edge computing mechanism was found beneficial to a great extent, and its effectiveness was well laid out. A major roadblock involving computational capabilities came to the fore while applying the edge computing model in LoRaWAN [2]. However, it has its advantages, which make its implementation quite beneficial. A smart city is an example of how technology has enhanced the day-to-day life of humans [3]. A smart city approach collects data from smart sensors using intelligent technologies to measure, analyze, store, and communicate with other systems [4]. The IoT (Internet of Things) is one of the major enabling factors for smart cities. In IoT, different devices

DOI: 10.1201/9781003218715-5

56

and nodes communicate via the Internet and are anticipated as efficient and cheap monitoring applications. To digitize the smart city, one of the most important criteria to be addressed is fading and localization [5]. IoT adds quality to services provided in a smart city and ensures that the sensor's information reaches the host; this is done by using specific communication protocols along with the infrastructure [6]. New technologies like Sigfox, NB-IoT and LoRaWAN facilitate long-distance wireless communication and are part of LPWAN families in which NB-IoT and LoRaWAN protocols are in trend more than protocols like Sigfox and Weightless [7].

Sigfox has an ultra-narrowband feature that allows the recipient to hear in a very less spectrum and is used to mitigate the noise, accomplished by long-range using slow MR (modulation rate). The major drawback of Sigfox over LoRaWAN is that users need to pay the charges after the deployment of Sigfox by network operators. In contrast, LoRa provides free usage of its independent network.

Weightless is introduced by the Weightless special interest group offering variations, Weightless-N and Weightless-P [8]. nWave communicates up to 3 km in only one direction transmission, which makes them unidirectional from base station (BS) to end users and features similar to Sigfox. On the other hand, Weightless-P uses GMSK and QPSK, offering bi-directional communication but can range up to 2 km and has shorter battery life. NB-IoT offers the flexibility of deployment using available bands in LTE. The communication occurs bidirectional OFDM, and SC-FDMA is used for downlink and uplink transmission. Compressing and optimizing the functionalities of LTE networks [8], NB-IoT is accomplished with low power consumption.

The power consumption of LoRa is less than NB-IoT, and thus the battery life of LoRa is greater, which makes it more feasible for smart cities as it requires refresh data. LPWAN can be considered the most prominent and effective approach for smart buildings because of its battery efficiency and bi-directional communication [9]. LoRaWAN promises the pervasive connectivity that smart city infrastructure requires.

LPWANs are one of the cutting-edge technologies which fulfill the requirement of low rate and long range in unlicensed sub-band GHz frequency bands and expected in the Fourth Industrial Revolution [4]. Figure 4.1 depicts the technologies features. It establishes a balanced network between long-range and low-power communications at the expense of throughput and is used in machine-to-machine (M2M) communication [10]. LoRa technology is one of the LPWAN platforms that provides the most promising and intelligent infrastructure to exchange the data between sensors and servers, the local network created by LoRa, acts as a host in the smart city management platform. In order to fulfill the demand of complex smart city applications, it requires multiple LPWAN nodes, which makes a high-capacity and scalable network of utmost importance. It also has monetary advantages wherein power lines can be used to provide power to the nodes. These power lines can be battery operated and reusable in most cases, thereby making them power efficient. However, smart city applications also have enhanced characteristics deemed important, including a diversity of devices with varied software platforms and widespread radio coverage [8].

Cities are diverse and congested, so line of sight is difficult to achieve. LoRa is the best approach as it works well in multipath channels. It also has good penetration and low path loss which endows full automation and is the best fit for implementing

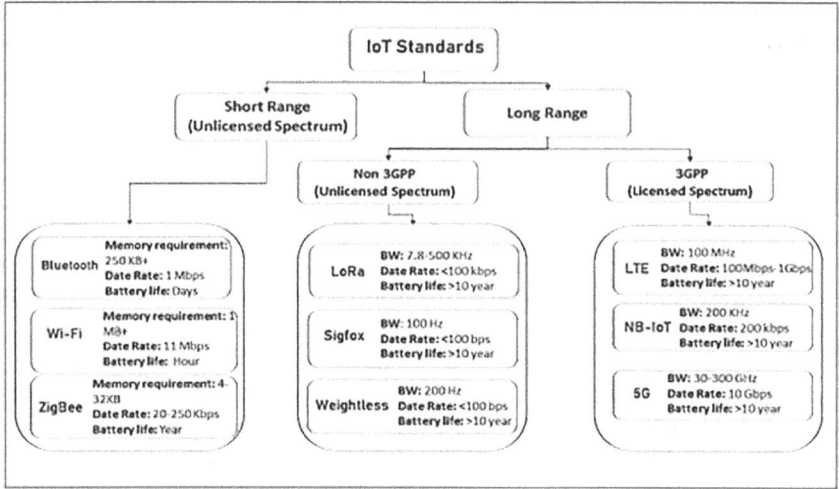

FIGURE 4.1   Technology classifications and features.

the smart city [11]. An open-source protocol is the most attractive feature that makes LoRa the best technology for smart cities or buildings. Low-power wide area networks are virtual interfaces between the industries and central servers where the data is collected, processed and stored [12].

New technological solutions are needed to optimize the increasingly scarce infrastructure resources, especially limited resources. Depending on the scalability and efficiency of smart cities, the most relevant studies and approaches are taken towards optimizing capacity. Channel capacity is the rate at which packets can be transmitted through a channel, and optimizing the capacity will increase the efficiency of the channel, and a smoother transition occurs. Different studies have been done in the field; below, some state of the art is discussed.

The author focused on smart waste management using advanced edge computing methodology using LoRaWAN class C. The five-layer architecture includes perception, network, middleware, application and finally, the business layer, where the network layer is depicted by LoRaWAN infra [2]. The other work showcases the power optimization on consumption focused on the end nodes and their distance from the gateway. The optimization is done with two different end devices, i.e., based on the distance normal and uniform distribution for the wider area and indoor, respectively [13]. In another work, the author focused on the limitations of the Indian network for various applications like street lighting, crop monitoring, and air pollution monitoring depending on gateway capacity on the length and rate of generation of messages. The technique analyzes the gateway capacity on the payload size of different applications and messages generated by end nodes [14]. Some of the work is highlighted in Table 4.1.

Existing works proposed methods to optimize the capacity by varying different parameters like spreading factor, code rate, bandwidth and payload of LoRa but failed to regard the most important aspects that parameters are simply varied, and

TABLE 4.1

Related Work and Their Approach

References	Study Purpose	Techniques	Outcomes
[2] A Multi-Layer LoRaWAN Infrastructure for Smart Waste Management.	LoRaWAN architecture for smart waste management	LoRaWAN classes are categorized, and a multi-layer structure is composed of different end nodes per complexity.	Optimize the process involved within waste management and track the filling level of drop-off containers and public bins.
[15] Optimal configuration of LoRa networks in smart cities.	Optimize the SFs and TPs assigned to each node to void collision and lower energy consumption	Integer linear programming model, OPT-MAX and OPT-DELTA	Optimized configuration performs consistently well, achieving a higher delivery ratio and a minimal energy consumption across different scenarios.
[4] Optimizing and Updating LoRa Communication Parameters: A Machine Learning Approach.	The accumulated average per-node throughput is maximized.	Evolution strategies (ES) algorithm	Remarkable increase in the accumulated average per-node throughput of 147% when the network is composed of 200 IoT nodes
[4] Performance optimization of LoRa nodes for the future smart city/industry.	The transmission policy C maximizes the throughput of the node while restricting power consumption to the maximum value of ω_{max} watts.	Genetic algorithm (GA) and simulated annealing (SA) algorithm	The goodness of our solution increases performance by more than 33.20%, 91.81% and 238.8% when compared to ADR, conservative and random policies, respectively.
[16] LoRaWAN for smart city IoT deployments: A long-term evaluation.	Deploying a city-scale LoRaWAN network across Southampton, UK, to support the installation of air quality monitors and to explore the capabilities of LoRaWAN	Deployment of custom-made gateways	On average, 72.4% of the messages sent were received, highlighting the need for alternative solutions when data completeness is required. 99% of the messages are received within 10 s of the transmission, which has implications for high-frequency sampling scenarios.
Current Work	Updating and optimizing channel capacity using different LoRa combinations for smart city applications	Particle swarm optimization algorithm	Critical comparative analysis of the optimized model with the non-optimized one reveals that ToA is reduced by approximately 15.66%, an observed improvement of 2.36% in channel capacity and 2.41% in SNR is observed.

its effects are being computed in terms of delivery ratio, power consumption, SNR and energy optimization. Channel capacity is not targeted directly, which is the most important parameter to be considered. The nodes and subscribers are numerous in smart cities and can directly affect the capacity and have an adverse effect on latency. However, literature related to LoRaWAN technology has increased in recent years.

Contributing and extending the above literature to acknowledge by segregating the constrained parameters, such as spreading factor, code rate and free parameters, i.e., bandwidth, carrier frequency, payload length, preamble length, and analytically observing the behavior of constrained parameters after the optimization using PSO. After several iterations, the best value of transmission time of smart nodes packet evaluated to reach the gateway is being reduced, thus improvising the channel capacity irrespective of spreading factors and code rate. Below we briefly address the LoRaWAN as a communication enabler for smart sensor nodes for various applications of smart cities and the work approach of this chapter.

The main contribution of the current work is:

- Analysis of smart city heterogeneous network is enabled by LoRa technology and comparison of purposed work concerning existing work.
- Evaluating the network performance presented the first mathematical LoRa model for smart application IoT nodes in terms of packet transmission time (Tpkt).
- Computational intelligence, i.e., the particle swarm optimization (PSO) technique, is introduced to optimize the IoT node performance and network efficiency. The algorithm is implemented to minimize node packets' transmission time or ToA and further analyze its effects on smart networks' SNR, BER, and channel capacity.
- A comparison of optimized networks and non-optimized networks is performed.
- The simulation result shows the comparative analysis of Tpkt (%), Channel capacity (%) and SNR (%), irrespective of spreading factor and coding rate, of highlighting the proposed PSO technique benefits.

4.2   TECHNOLOGY PRELIMINARIES IN TERMS OF IOT

Currently, the communication technology sector is rapidly gaining momentum, which leads to its expanding boom in various applications of IoT. Thus, an increasing number of pertinence has specific requirements, like the range needs to be as wide as possible to cover more area in the limited frame. Data rate should be minimal to maximize the usable distance, the potential of minimal energy consumption and cost-effectiveness to justify the particular technology with potential. Low power wide area network (LPWAN) advancement in both licensed and non-licensed spectrum leads to the emergence of many technologies like Sigfox, LoRa and NB-IoT with various technical differences. Each technologies feasibility depends on its factors and specific applications such as:

Deployment Structure: LoRa has its prowess over Sigfox and NB-IoT, as it offers local implementation by connecting public networks via BS (base stations) and LAN using LoRa gateway. In industries, LoRa deployment can be achieved by the hybrid model using a local LoRa network, and outside areas are covered using a public network.

Range and Coverage: Contrary to LoRa and Sigfox, NB-IoT is constrained to LTE base station, whereas coverage is less than 10 km. Thus the region with no LTE coverage, NB-IoT is not suitable in rural or suburban areas. On the other hand, Sigfox and LoRa have the advantage over this. They cover wider areas, i.e., 20–30 km range with two or three base stations [17].

Cost: Technology growth comes with different aspects like license (spectrum), deployment, hardware, etc. In LoRa and Sigfox, data transmission occurred on an unlicensed spectrum. LoRa and Sigfox, compared to NB-IoT, is more cost-effective.

Scalability: The technology must connect substantial users and devices and work well with a progressive number of end devices. On the contrary, NB-IoT is much more efficient for scalability features. NB-IoT can add a large number of end devices per cell, i.e., approximately 1 lakh, whereas Sigfox and LoRa can occupy 50,000 users in one cell. In addition to this, NB-IoT offers a maximum payload and can transmit data up to 1.6k bytes. LoRaWAN can offer up to 243 bytes. Sigfox offers 12 bytes for uplink data and 8 bytes for downlink data, limiting the use of technology for various IoT applications.

4.2.1 BATTERY LIFE

Technology must use less power and consume minimal energy to increase the battery life. In LoRa, Sigfox and NB-IoT, the end devices can have a long lifetime. For any IoT applications, the device must last long and to optimize the power. The devices must be in sleep mode when not in use. Whereas NB-IoT comparatively consumes additional power because of synchronous communication and the multiplexing technique, it involves OFDM (orthogonal frequency division multiplexing) requiring peak current.

4.2.2 LATENCY

Latency is the most important criterion when the application is sensitive. A delay in data transmission can be acceptable when the application is insensitive. Thus, for minimal latency, NB-IoT and Class C LoRa WAN is the appropriate technique to be considered. Where fewer data and delay insensitive transmission is necessary [17], then class A LoRa and Sigfox can be considered.

4.2.3 QoS

Quality of service (Qos) is a factor that manages the resources and priorities of specific data on the network by minimizing packet loss and jitter. Sigfox and LoRa work on unlicensed ISM spectrum, whereas NB-IoT works on licensed spectrum and is preferred for service constrained quality [17].

4.3   LPWAN AND ITS APPLICATION IN SMART CITY

Smart cities use different technologies to improvise existing services, which helps enhance people's quality of life. Technology is developing to realize smart cities, which are referred to as the engine of economic growth [4]. LPWAN is an advanced version of communication networks focused on battery-powered and resource-constricted IoT devices and a solution that depends on the LoRa physical layer for long-range connectivity at low data rates [18]. The major reasons like energy efficiency, multipath channel propagation, and bi-direction communication make the LoRaWAN technology is best suitable for smart city applications because:

1. The most required parameter is SECURITY, and LoRa provides the security on both the layers, i.e., the network and application layers.
2. It is less costly comparing the other licensed spectrum-based technologies as LoRa is open source and unlicensed spectrum.
3. Energy efficiency and battery life are the ideal parameters as the transmission between networks occurs only when required, which helps to save the battery and reduce energy consumption and results in longer duration sustainability.
4. It includes multipath channel propagation, as LoRa easily receives the signal penetrated from the walls or elevator shaft, thus adding the features of good penetration and low path loss.

Smart city scenarios have no restricted QoS requirements and occasionally send a small amount of data supported by data rates of LoRa [11].

4.3.1   LPWAN Technology Overview

LPWAN network structure is divided into two parts: master and slave functions. The end devices, functional IoT devices, are slaves, and the network server acts as a master. The network manages the alteration in the slave or the functionality change required in the slave [19]. LoRa and LoRaWAN are contradictory. Combining those makes LPWAN. LoRa stands for long range, and patent by Semtech Corporation, which exists in the physical layer, is an infrastructure-less modulation technology that helps in transferring information over a large area having a small value of the transfer rate enabling long-range communication link [7]. LoRaWAN, on the other hand, exists in the data link layer and MAC sub-layer, which majorly influences the battery lifetime, QoS, and network capacity. The most interesting fact that makes the LoRaWAN empowered technologies, among others, is the ability of nodes over long-range distances to transmit data. In an energy-constrained device, we prefer to use as less power usage for longer distance transmission by the modulation technique, which involves increasing receiver sensitivity so that the Rx can demodulate the data with low power and thus spread spectrum techniques to accomplish this benchmark.

TABLE 4.2
Frequency Band Based on Operative Region (MHz) [7]

Europe	North America	China	Japan	India	Australia
433 and 863–870	902–928	470–510 and 779–787	920–923	865–867	915–928

LoRa falls under the spread spectrum (SS) modulation category. The best part of this technique is that it uses the chirp signal method, in which the frequency alters steadily for the transmission of the information. This method has the upper hand over the other methods, as both the transmitter and receiver have the same offset time and frequency, which cuts down the receiver's complication level [4].

Mainly LoRaWAN works in an unlicensed ISM band, i.e., depending on the region of operation, making it cost less. Table 4.2 shows access bands respective to the regions.

LPWAN technology comprises the LoRa physical and MAC layer, in which the LoRa physical layer avoids noise interference and makes robust communication which also helps to increase receiver sensitivity. Every information bit is represented by multiple chips of information transmitted over the communication channel bandwidth between 7.8 kHz to 500 kHz [12]. The spreading factor (SF) is the number of chirps per symbol; SF increases lead to the increment of chirps per symbol, which also concludes with a decrement in data rate. Further section contributes to the smart city model scenario where LoRa technology is channelized, and various parameters are being mathematically modeled for different configurations.

4.4 SYSTEM MODEL

This section focuses on channel capacity evaluation and other LoRa WAN parameters for smart applications like farming, education and health services. The capacity and efficiency should be higher. Topologies are the basic structure of the network, and the most prominent are star-on-star and mesh. Comparatively, star-on-star topologies are preferred over mesh as the power consumption is increased due to more sleep state and communication range [8]. LoRa star topology connectivity facilitates a single-hop network to substantially a huge number of nodes, making them fast and reliable. One gateway node is linked to other end devices and can communicate via gateway only. The gateway further transmits the received data from multiple recipients to a network server, where errors and redundancy are verified. If the nodes malfunction, it can be easily traceable and repaired without affecting the other connectivity and transmission.

LoRa mesh topology, multiple end nodes or sensor nodes can transmit messages via various paths to router nodes, which further transmit it to the gateway node and thus increase transmission reliability. The disadvantage of mesh topology is the cost and complexity, which will proportionally increase as the network expand.

The scenario is shown in Figure 4.2. It is assumed that N number of smart devices were deployed and enabled by LoRa technology, and network communication

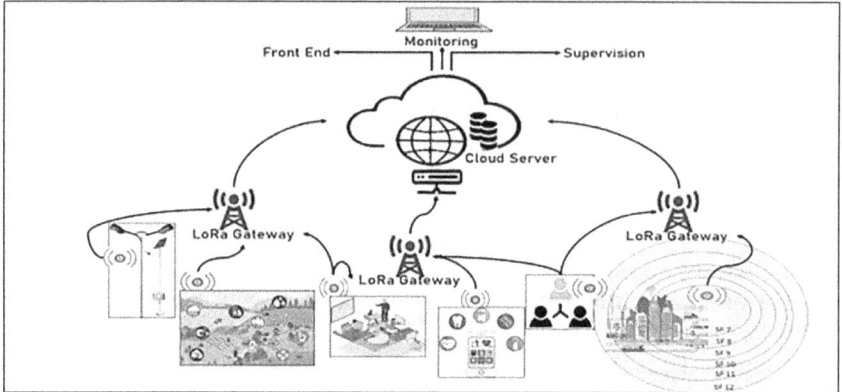

FIGURE 4.2   LoRa architecture for smart city applications.

channels are allocated depending on the regional parameters monitored by gateway devices and respectively configured. IoT sensors have been deployed for smart applications. The reason behind this is to extend the network's lifespan as nodes remain in sleep mode most of the time. After a periodic interval, nodes came into the wake-up state, sensed the physical parameters, and reported to the LoRa gateway as necessary.

Sensors sensed and processed the data from various intelligent nodes and sent it toward the sink node. The sink node further transmitted it towards the LoRa gateway in the respective channel and bandwidth of the region. The smart sensors are in cycle mode, i.e., sleep mode and periodicity of T seconds to wake up and sense the physical parameters. If data is valid, it sends to the LoRa gateway, which is always active. In smart city applications, there is supposed to be a large variety of phenomena monitored; thus, according to the priorities, data are processed. During communication, channels are dedicated to their respective processes, i.e., the "main channel" is dedicated to data transmission.

In contrast, LoRa Gateway responses to smart nodes (LoRa) are dedicated to the "downlink channel"; the "uplink channel" is used to send the request towards the gateway from smart nodes [20]. These data are sent to the cloud server, monitored, supervised, and stored. During packet transmission (Tpkt), LoRa nodes randomly select the channel and transmit it towards the gateway. The packet's length (Lpkt) varies depending on the LoRa parameter settings.

The main goal is to achieve a smart city model with high performance and maximum capacity. The model is achieved and simulated by customizing the LoRa parameters, i.e., Carrier Frequency, Coding Rate, Bandwidth, and Spreading Factor. Table 4.3 introduces the LoRa Model configured parameters. Spreading factors are orthogonal (which also helps in network separation) and range from 7 to 12. Also can be a trade-off between data rate and coverage range.

Spreading factors are orthogonal to each other (which also helps in network separation) and range from 7 to 12; also can be a trade-off between data rate and coverage range—the greater the SF, the greater the sensitivity, leading to high transmission time. To overcome the burst interferences and robust communication, LoRa uses FEC (forward error correction) with code rates between 4/8 to 4/5; the higher the

TABLE 4.3

LoRa Parameters

Parameters	Values
BW (Communication Channel Bandwidth)	125kHz, 250kHz, 500kHz
Carrier Frequency	865 MHz
SF (Spreading Factor)	7,10,12
CR (Code Rate)	4/5, 4/6, 4/7, 4/8
PL (Payload in bytes)	8
n_p (Programmed Preamble in Symbols)	6
P_t (Transmit Power in dBm)	17
H (Header)	0
DE (Data Rate)	0 {Else}
	1 {SF-11,12 at BW-125kHz)
Duty Cycle (in ms)	2000

code rate, the more robustness in communication which also cost increases in transmission time. The spreading factor and code rate were optimized to increase the network's performance by evaluating the transmission time. Channel capacity and SNR of the network are dependent on packet transmission time.

In LoRa network, the chirp signal carries baseband data and the frequency range of particular chirp is bandwidth, therefore the chip rate (R_c) is

$$Rc = BW \frac{chips}{sec} \tag{4.1}$$

The SF (spreading factor) relate symbols and chips. Symbol rate (R_s) is the rate at which the spread information is sent

$$Rs = R_c / 2^{SF} \tag{4.2}$$

In LoRa network, above discussed that to avoid the congestion and interference, forward error correction is added in every T_{pkt}. Increasing CR (code rate) helps to avoid increased interference in channel and leads to decrease in transmission time (T_{pkt}).

$$R_b = SF * \frac{4}{4+CR} * R_s \tag{4.3}$$

On contrary of equation of R_c, R_b and Rs [9], let us assume that p packets are sent and T_{pkt} is the transmission time taken by node for transmitting LoRa packets, which is given as

$$T_{pkt} = T_{preamble} + T_{payload} \tag{4.4}$$

LoRa network packet is composed by a preamble which is for synchronization, payload and respective data fields. Therefore, transmission time of LoRa node is given by the following equation below:

$$T_{pkt} = \left\{ \left(n_p + 4.25 \right) 1 / R_s \right\} +$$

$$\left\{ 8 + max \left[ceil \left(\frac{8PL - 4\left(log_2 R_c / R_s \right) + 28 + 16 - 20H}{4\left(log_2 \left(R_c / R_s \right) - 2DE \right)} \right), (CR+4), 0 \right] \right\} 1 / R_s \quad (4.5)$$

Percentage difference of optimized and non-optimized Tpkt is given in equation 6, where Tpkt NO and Tpkt O are transmission time non optimized and optimized respectively. TOA (%) has been observed to show significant improvement which helps to reduce the latency.

$$TOA(\%) = \left[\left(T_{pkt}O - T_{pkt}NO \right) / \left(\frac{T_{pkt}O + T_{pkt}NO}{2} \right) \right] * 100 \quad (4.6)$$

Since transmission time is the duration of packet reaching destination, and discussed previously that the length of the packet is variable and given as:

$$L_{pkt} = \left\{ \left(n_p + 4.25 \right) 1 / R_s \right\} +$$

$$\left\{ 8 + max \left[ceil \left(\frac{8PL - 4\left(log_2 R_c / R_s \right) + 28 + 16 - 20H}{4\left(log_2 \left(R_c / R_s \right) - 2DE \right)} \right), (CR+4), 0 \right] \right\} 1 / R_s + R_b \quad (4.7)$$

SNR and BER is the most important metrics which helps to know the link quality and further with the help of this capacity evaluation is done.

$$BER = 1 / \frac{E_b}{N_0} * \left(\frac{\left\{ \left(n_p + 4.25 \right) 1 / R_s \right\} + \left\{ 8 + max \left[ceil \left(\frac{8PL - 4\left(log_2 R_c / R_s \right) + 28 + 16 - 20H}{4\left(log_2 \left(R_c / R_s \right) - 2DE \right)} \right), (CR+4), 0 \right] \right\} 1 / R_s + R_b}{\left\{ \left(n_p + 4.25 \right) 1 / R_s \right\} + \left\{ 8 + max \left[ceil \left(\frac{8PL - 4\left(log_2 R_c / R_s \right) + 28 + 16 - 20H}{4\left(log_2 \left(R_c / R_s \right) - 2DE \right)} \right), (CR+4), 0 \right] \right\} 1 / R_s} \right)^K \quad (4.8)$$

The signal to noise ratio (SNR) of the LoRa communication is measured with the help of the equation

$$SNR = \frac{E_b}{N_0} * \frac{\left\{ \left(n_p + 4.25 \right) 1 / R_s \right\} + \left\{ 8 + max \left[ceil \left(\frac{8PL - 4\left(log_2 R_c / R_s \right) + 28 + 16 - 20H}{4\left(log_2 \left(R_c / R_s \right) - 2DE \right)} \right), (CR+4), 0 \right] \right\} 1 / R_s + R_b}{\left\{ \left(n_p + 4.25 \right) 1 / R_s \right\} + \left\{ 8 + max \left[ceil \left(\frac{8PL - 4\left(log_2 R_c / R_s \right) + 28 + 16 - 20H}{4\left(log_2 \left(R_c / R_s \right) - 2DE \right)} \right), (CR+4), 0 \right] \right\} 1 / R_s} \quad (4.9)$$

Percentage difference of optimized and non-optimized SNR is given in equation 4.10, where SNR_O and SNR_{NO} are Signal to Noise ratio at non optimized and optimized achieved respectively.

$$SNR(\%) = \left[\left(SNR_O - SNR_{NO} \right) / \left(\frac{SNR_O + SNR_{NO}}{2} \right) \right] * 100 \qquad (4.10)$$

Capacity or network capacity refers to how much gateway receive messages at a time from different nodes, achieved with multi-channel Tx and AER, i.e., adaptive data rate (optimize data rate and energy consumption). Increased channel capacity leads to addition of more smart nodes for actively participation in network.

$$C = R_c * log_2 \left(1 + \frac{E_b}{N_0} * \frac{\left\{ (n_p + 4.25)1/R_s \right\} + \left\{ 8 + max \left[ceil \left(\frac{8PL - 4(log_2 R_c / R_s) + 28 + 16 - 20H}{4(log_2 (R_c / R_s) - 2DE)} \right), (CR+4), 0 \right] 1/R_s + R_b \right\}}{\left\{ (n_p + 4.25)1/R_s \right\} + \left\{ 8 + max \left[ceil \left(\frac{8PL - 4(log_2 R_c / R_s) + 28 + 16 - 20H}{4(log_2 (R_c / R_s) - 2DE)} \right), (CR+4), 0 \right] 1/R_s \right\}} \right) \qquad (4.11)$$

Percentage difference of optimized and non-optimized channel capacity is given in equation 4.12, where C_O and C_{NO} are capacity of channel at non optimized and optimized achieved, respectively.

$$C(\%) = \left[\left(C_O - C_{NO} \right) / \left(\frac{C_O + C_{NO}}{2} \right) \right] * 100 \qquad (4.12)$$

In further sections, configured profile with different settings are integrated in simulation models along with optimized and updated data achieved by the PSO [21] technique introduced for increasing efficiency of network by increasing the capacity and SNR.

4.5 PERFORMANCE ANALYSIS AND SIMULATED RESULT

This section includes the optimization technique PSO (particle swarm optimization), ensuring the transmission packet duration (Tpkt) will be improvised. Tpkt depends on various parameters that need to be separated as free and constrained. Tpkt behavior will be analyzed based on constrained parameters.

The number of end devices will increase as smart nodes sense the data and transmit it to the LoRa gateway. Tpkt is minimum in SF-7, as the nearer region takes minimum time to transmit the packet to the nodes. Since CR (code rate) increases, the time duration of transmission (ToA) decreases and thus will achieve a minimum at CR: 4/5 in the communication channel. Table 4.4 shows the dataset value of optimized T_{pkt} achieved with the help of PSO.

From the equations in the above section conclude transmission packet duration is dependent on R_b, R_c, R_s and CR, which are constrained, and np, PL, H and DE, which

TABLE 4.4

Optimized Tpkt Achieved Combination

BW (Bandwidth) (kHz)	CR (Code Rate)	SF (Spreading Factor)	Non-Optimized Tpkt (ms)	Optimized Tpkt (ms)
125	4/5	7	34.05	33.37
	4/5	10	231.42	219.14
	4/5	12	761.86	562.82
250	4/5	7	17.02	16.04
	4/5	10	115.71	93.01
	4/5	12	380.93	368.33
500	**4/5**	**7**	**8.51**	**8.42**
	4/5	10	57.86	55.50
	4/5	12	190.46	175.77

are free parameters. From the equations constrained parameters, i.e., R_b, R_c and R_s are dependent on spreading factor (SF), bandwidth (BW) and code rate (CR) in which SF and CR are simultaneously taken as constrained, which in turn optimize the Tpkt, which generate random variables and optimization is to choose the best fit among other solutions (i.e., to choose the best z from a set of solution Z). Optimization will be accomplished with the help of PSO.

> *Particle swarm optimization*: Optimization is necessary to increase the efficiency to an optimum level and yet be cost-effective. Thus, genetic algorithm and PSO are used to increase the network efficiency as much as possible to a certain level. The algorithm is quite similar, but the PSO is more cost-effective and less iterations used [22].

A genetic algorithm was introduced in the 1970s by John Holland. It usually accommodates biological techniques like mutation, selection or inheritance. GA can resolve complex optimization problems as it caters to continuous and discrete variables. PSO was introduced in the 1990s [21], and the idea is that each particle depicts a solution which will be updated by cognitive and social behavior. The particle swarm optimization algorithm takes random values based on the current position [22] in particle space and equivalent velocity vector. A genetic algorithm could not deal with complexity adequately as search space will expand when the large number of elements going through the mutation, and thus PSO are considered for more complexity as parameters are less, also take a smaller number of iterations.

PSO is an algorithm based on population-based searching which works in a multi-dimensional search space in which particles fly with adaptable velocities.

It works by generating random numbers of a solution, and each solution is iteratively updated and provides two values: local best and global best. Based on these particles, velocities are along with its position update, and fitness is calculated of that particle. This way, an optimal solution is achieved after a certain iteration is processed. The efficiency of PSO is influenced by the quality and quantity of numbers

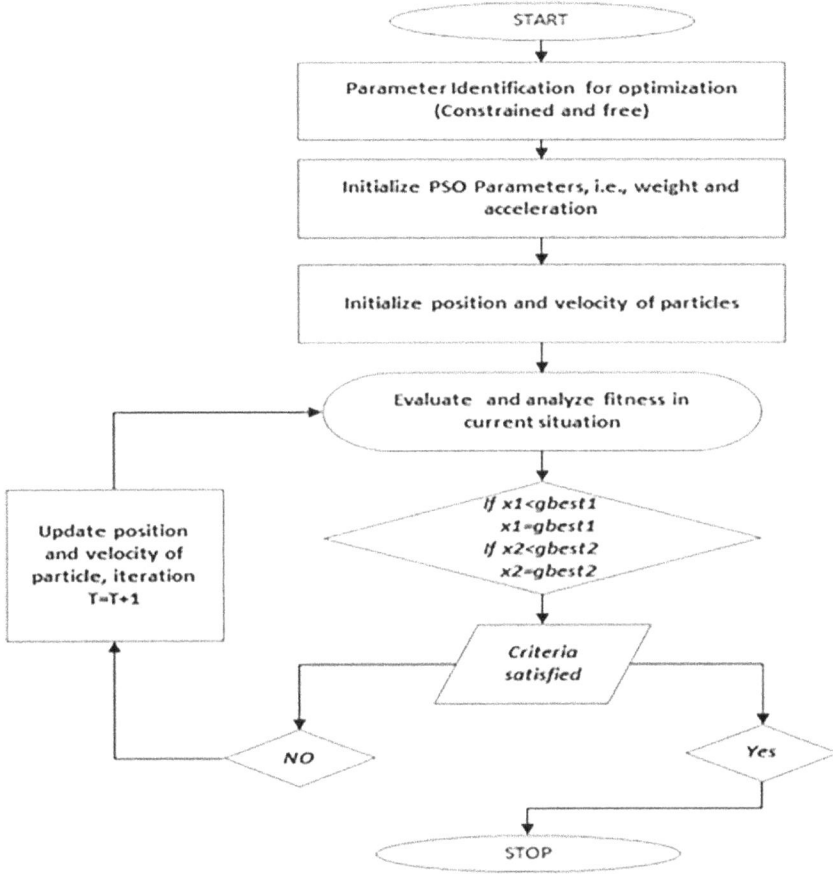

FIGURE 4.3 PSO algorithm flow.

accomplished. PSO converges towards global optima, whereas GA focuses on local optimum [22]. PSO workflow is discussed in Figure 4.3 [23] along with the parameter values for achieving the optimal solution in Table 4.5.

The simulation is based on a mathematical model of LoRa communication introduced in section 4.3, which is capable of optimizing and maximize the capacity along with improving the SNR with a decrease in Tpkt with the help of optimizing technique introduced in section 4.4. As increasing the spreading factor, the transmission time of smart node packets increases.

TABLE 4.5
PSO Parameters

Iteration Maximum	100
Initial Weight (w_{max})	0.9
Final Weight (w_{min})	0.4
Acceleration (c1,c2)	1.4455

Comparative analysis of optimized and non-optimized ToA is observed to be reduced by 15.66% irrespective of different SF, i.e., 7, 10 and 12, respectively for BW-125 kHz, 250 kHz and 500 kHz in Figures 4.4–4.6. The minimum transmission time is achievable at 500 kHz. Length of the packet (Lpkt) depends on the parameters Rb, Rc, Rs and CR, which are constrained, and np, PL, H and DE are mentioned in Table 4.3. Packet length optimization is critical issues in WSN for improving network performance, i.e., network lifetime and reliability as longer packet length leads to an increase in high loss rates whereas shorter packet length encounters greater overhead. Therefore, the optimal value of Lpkt is chosen with a different parameter setting of LoRa. Table 4.6 depicts the data value of the combination at which Tpkt achieved is low.

Since the transmission time should be minimum to get the best performance and capacity will be maximized as the capacity evaluation is essential for the proper functioning of IoT. Further, SNR of the network is dependent on the length of packet and transmission time of packets from LoRa nodes towards gateway SNR improvement leads to improvisation of link quality. The SNR relationship in

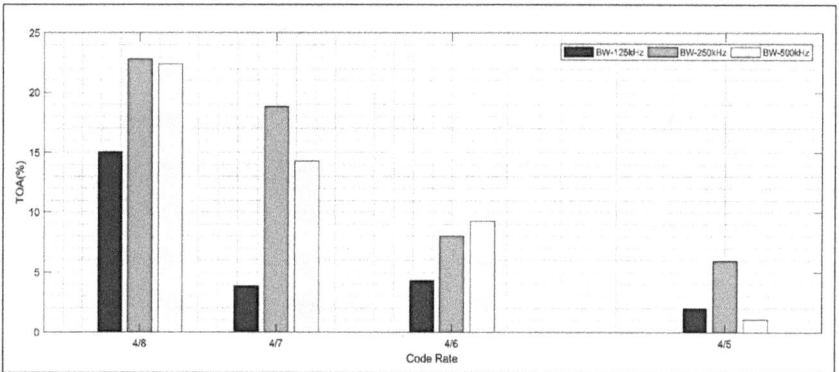

FIGURE 4.4   Percentage difference of TOA for different bandwidth at SF-7.

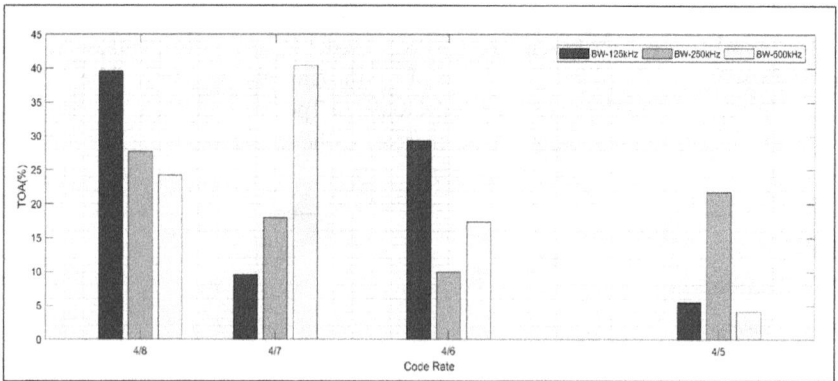

FIGURE 4.5   Percentage difference of TOA for different bandwidth at SF-10.

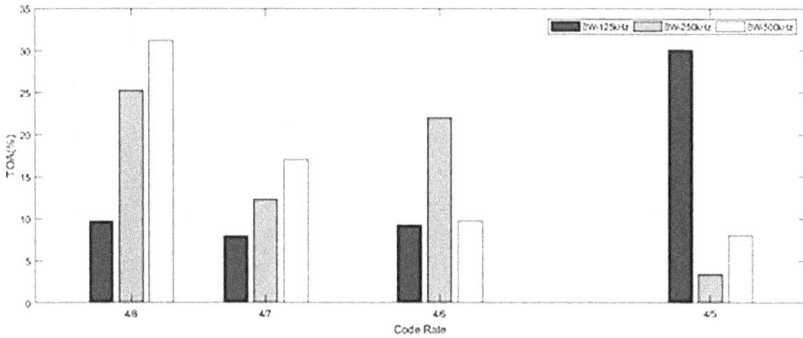

FIGURE 4.6 Percentage difference of TOA for different bandwidth at SF-12.

TABLE 4.6
Optimized Lpkt Achieved Combinations

BW (Bandwidth) (kHz)	CR (Code Rate)	SF (Spreading Factor)	Non-Optimized Lpkt	Optimized Lpkt
125	4/5	7	5.5028×10^3	5.5021×10^3
	4/5	10	1.20×10^3	1.19×10^3
	4/5	12	1.0548×10^3	8.5579×10^2
250	4/5	7	1.0955×10^4	1.0954×10^4
	4/5	10	2.06×10^3	2.04×10^3
	4/5	12	9.668×10^2	9.5427×10^2
500	4/5	7	2.1884×10^4	2.1883×10^4
	4/5	10	3.96×10^3	3.396×10^3
	4/5	12	1.3623×10^3	1.3477×10^3

equation shows that the minimum transmission time of the node packet to reach the gateway leads to the best SNR value. Table 4.7 shows the maximum SNR achieved is at 500 kHz compared to 125 kHz and 250 kHz where increasing CR leads to get better SNR and thus, at CR-4/5 minimized Tpkt and maximum signal strength is found.

The simulated graphs in Figures 4.7–4.9, depicting the percentage difference of the optimized value of SNR in comparison to non-optimized data, show the improvement of average approximately 2.41% irrespective of spreading factor and code rate for all bandwidths.

Capacity plays an important role in the network smooth flow and ideally, the capacity of the network is considered desirable when it increases and leads to no interference and congestion in the network. Practically no interference and congestion opted is not possible, but optimizing the network helps to get the minimum error or congestion which can help to increase the capacity to maximum level where

TABLE 4.7

Maximum SNR Achieved Combinations

BW (Bandwidth) (KHz)	CR (Code Rate)	SF (Spreading Factor)	Non-Optimized SNRmax (dB)	Optimized SNRmax (dB)
125	4/5	7	16.43	16.48
	4/5	10	8.98	9.08
	4/5	12	6.10	6.30
250	4/5	7	19.43	19.56
	4/5	10	11.65	12.10
	4/5	12	7.41	7.46
500	4/5	7	22.44	22.46
	4/5	10	14.57	14.66
	4/5	12	9.66	9.81

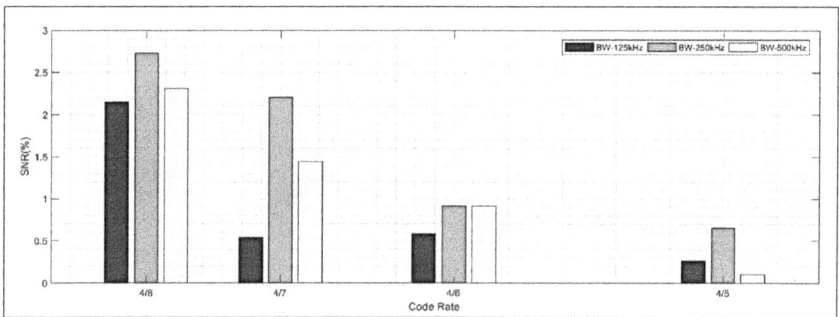

FIGURE 4.7   Percentage difference of SNR for different bandwidth at SF-7.

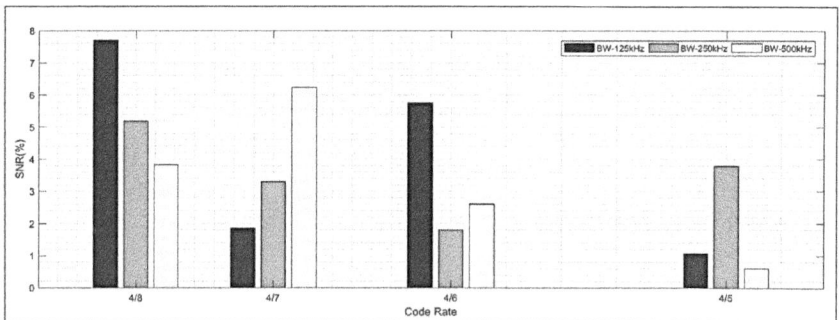

FIGURE 4.8   Percentage difference of SNR for different bandwidth at SF-10.

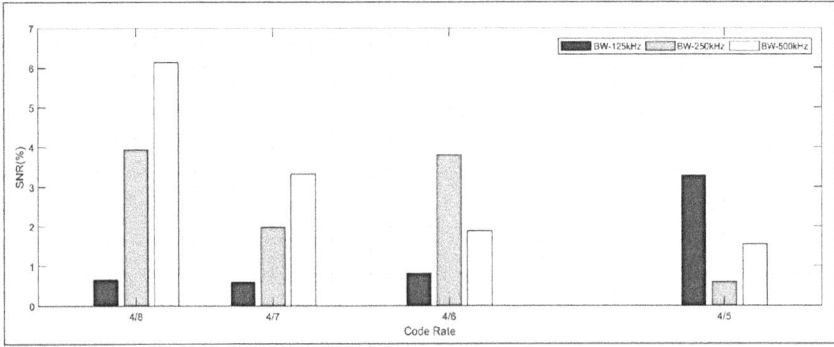

FIGURE 4.9   Percentage difference of SNR for different bandwidth at SF-12.

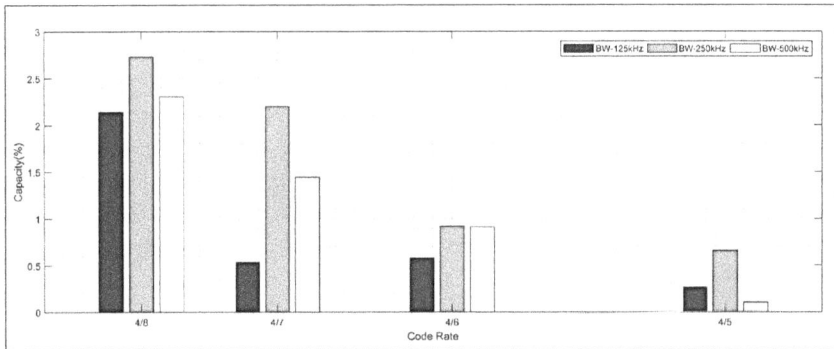

FIGURE 4.10   Percentage difference of capacity for different bandwidth at SF-7.

network performance can be best achieved, and more nodes can be accommodated in the network. Capacity I depends on the parameters Rc, Eb/N0, Lpkt and Tpkt, which are constrained and achieved above.

In Figures 4.10–4.12, the simulated result shows that channel capacity is improved averagely by 2.36% overall irrespective of LoRa parameters, i.e., spreading factor and code rate.

Table 4.8 depicts that when the transmission channel is 500 kHz maximum, capacity is achieved at this configuration at SF (spreading factor) at 7 and minimum is at SF-12, when the bandwidth is 125 kHz. Thus with increasing Tpkt the capacity decreases. The table also shows the maximum capacity of the channel which is achieved, and the optimum value achieved by non-optimized and optimized Tpkt. Thus, with the simulated result and parameter combination, the maximum channel capacity is achieved at 500 kHz, at SF-7 when the code rate is 4/5 shown with highlight.

TABLE 4.8

Maximum Channel Capacity Achieved Combinations

BW (Bandwidth) (kHz)	CR (Code Rate)	SF (Spreading Factor)	Non-Optimized Capacity$_{max}$ (bits/sec/kHz)	Optimized Capacity$_{max}$ (bits/sec/kHz)
125	4/5	7	1.3653×10^6	1.3689×10^6
	4/5	10	7.4898×10^4	7.5685×10^5
	4/5	12	5.1734×10^5	5.333×10^5
250	4/5	7	3.2288×10^6	3.2502×10^6
	4/5	10	1.9380×10^6	2.0124×10^6
	4/5	12	1.2438×10^6	1.2510×10^6
500	4/5	7	7.4567×10^6	7.4644×10^6
	4/5	10	4.8425×10^6	4.8721×10^6
	4/5	12	3.2201×10^6	$3.2696\ 0^6$

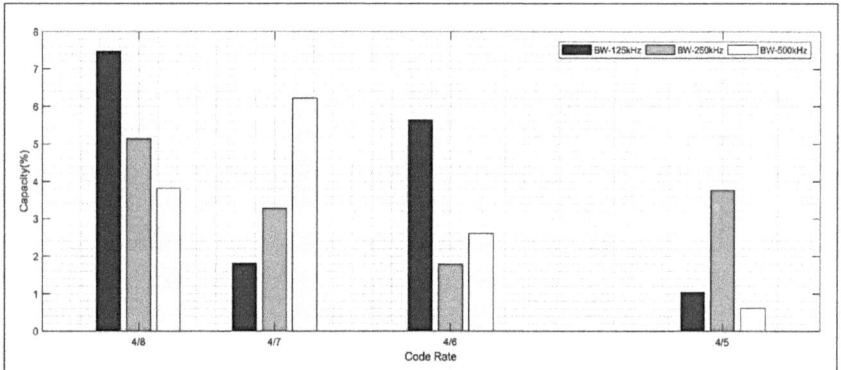

FIGURE 4.11   Percentage difference of capacity for different bandwidth at SF-10.

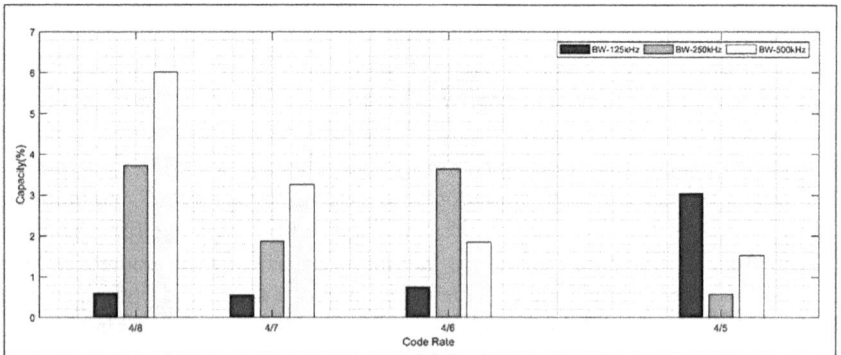

FIGURE 4.12   Percentage difference of capacity for different bandwidth at SF-12.

4.6 CONCLUSION AND FUTURE PERSPECTIVES

This chapter focuses on applying and optimizing LoRa and LoRaWAN for innovative city applications. A mathematical model to evaluate Tpkt, SNR and BER capacity has been formulated. Compared to other licensed or unlicensed spectrum-based technologies, LoRa has advantages.

It has been shown that different LoRa parameter configuration is being tested, and the best setting is chosen for the appropriate transmission. Channel capacity and its related parameters are seen to be dependent on various free and constrained parameters. The Tpkt has been optimized using the PSO algorithm. Further, using the optimized value of Tpkt Channel capacity of the proposed model has been optimized by getting the optimal solution of transmission time (Tpkt) of smart node packets to reach the LoRa gateway. Simulation results reveal that in the considered model, ToA, irrespective of the spreading factor and code rate after optimization, is approximately improved 15.66%, decrement in ToA indicates minimized latency in the smart city as the data packets will travel fast from the subscriber nodes towards the gateway node. Optimization improves channel capacity by 2.36%. Increase channel capacity so more nodes can be deployed in a smart city. SNR also shows an increment of 2.41%, indicating an improvement in the link quality of the network. The future scope lies in improvising smart nodes' power management and energy efficiency by making a more effective deployment strategy and planning. This can be achieved by optimizing sensor nodes' power, latency and coverage.

REFERENCES

[1]. Punt, E. P., Geertman, S. C. M., Afrooz, A. E., Witte, P. A., & Pettit, C. J. (2020). Life is a scene and we are the actors: Assessing the usefulness of planning support theatres for smart city planning. Computers, Environment and Urban Systems, 82, 101485.

[2]. Baldo, D., Mecocci, A., Parrino, S., Peruzzi, G., & Pozzebon, A. (2021). A multi-layer LoRaWAN infrastructure for smart waste management. Sensors, 21(8), 2600. DOI: 10.3390/s21082600

[3]. Mayaud, J. R., Tran, M., Pereira, R. H., & Nuttall, R. (2019). Future access to essential services in a growing smart city: The case of Surrey, British Columbia. Computers, Environment and Urban Systems, 73, 1–15.

[4]. Sandoval, R. M., Garcia-Sanchez, A. J., & Garcia-Haro, J. (2019). Performance optimization of LoRa nodes for the future smart city/industry. EURASIP Journal on Wireless Communications and Networking, 2019(1), 1–13. DOI: 10.1186/s13638-019-1522-1

[5]. Asadullah, M., & Celik, A. R. (2016). An effective approach to build smart building based on internet of things (IoT). Journal of Basic and Applied Scientific Research (JBASR), 6(5), 56–62.

[6]. Van den Abeele, F., Haxhibeqiri, J., Moerman, I., & Hoebeke, J. (2017). Scalability analysis of large-scale LoRaWAN networks in ns-3. IEEE Internet of Things Journal, 4(6), 2186–2198. DOI: 10.1109/JIOT.2017.2768498.

[7]. Ayoub, W., Samhat, A. E., Nouvel, F., Mroue, M., & Prévotet, J. C. (2018). Internet of mobile things: Overview of LoRaWAN, DASH7, and NB-IoT in LPWANs standards and supported mobility. IEEE Communications Surveys & Tutorials, 21(2), 1561–1581. DOI: 10.1109/COMST.2018.2877382

[8]. Chaudhari, B. S., Zennaro, M., & Borkar, S. (2020). LPWAN technologies: Emerging application characteristics, requirements, and design considerations. Future Internet, 12(3), 46. DOI: 10.3390/fi12030046

[9]. Raza, U., Kulkarni, P., & Sooriyabandara, M. (2017). Low power wide area networks: An overview. IEEE Communications Surveys & Tutorials, 19(2), 855–873. DOI: 10.1109/COMST.2017.2652320

[10]. Bor, M. C., Roedig, U., Voigt, T., & Alonso, J. M. (2016, November). Do LoRa low-power wide-area networks scale? In Proceedings of the 19th ACM International Conference on Modeling, Analysis and Simulation of Wireless and Mobile Systems (pp. 59–67). ACM. DOI: 10.1145/2988287.2989163

[11]. Curry, E., Dustdar, S., Sheng, Q. Z., & Sheth, A. (2016). Smart cities–enabling services and applications. Journal of Internet Services and Applications. DOI: 10.1186/s13174-016-0048-6

[12]. Loriot, M., Aljer, A., & Shahrour, I. (2017, September). Analysis of the use of LoRaWan technology in a large-scale smart city demonstrator. In 2017 Sensors Networks Smart and Emerging Technologies (SENSET) (pp. 1–4). IEEE. DOI: 10.1109/SENSET.2017.8125011

[13]. Fialho, V., & Fortes, F. (2019, September). Low power IoT network sensors optimization for smart cities applications. In 2019 International Conference on Smart Energy Systems and Technologies (SEST) (pp. 1–6). IEEE. DOI: 10.1109/SEST.2019.8849071

[14]. Dalela, P. K., Sachdev, S., & Tyagi, V. (2019, March). LoRaWAN network capacity for practical network planning in India. In 2019 URSI Asia-Pacific Radio Science Conference (AP-RASC) (pp. 1–4). IEEE. DOI: 10.23919/URSIAP-RASC.2019.8738342

[15]. Premsankar, G., Ghaddar, B., Slabicki, M., & Di Francesco, M. (2020). Optimal configuration of LoRa networks in smart cities. IEEE Transactions on Industrial Informatics, 16(12), 7243–7254.

[16]. Basford, P. J., Bulot, F. M., Apetroaie-Cristea, M., Cox, S. J., & Ossont, S. J. (2020). LoRaWAN for smart city IoT deployments: A long term evaluation. Sensors, 20(3), 648. DOI: 10.3390/s20030648

[17]. Mekki, K., Bajic, E., Chaxel, F., & Meyer, F. (2019). A comparative study of LPWAN technologies for large-scale IoT deployment. ICT Express, 5(1), 1–7. DOI: 10.1016/j.icte.2017.12.005

[18]. Luvisotto, M., Tramarin, F., Vangelista, L., & Vitturi, S. (2018). On the use of LoRaWAN for indoor industrial IoT applications. Wireless Communications and Mobile Computing, 2018.

[19]. Kim, D. Y., Kim, S., & Park, J. H. (2020). A combined network control approach for the edge cloud and LPWAN-based IoT services. Concurrency and Computation: Practice and Experience, 32(1), e4406. DOI: 10.1002/cpe.4406

[20]. Paul, B. (2020). A novel mathematical model to evaluate the impact of packet retransmissions in LoRaWAN. IEEE Sensors Letters, 4(5), 1–4. DOI: 10.1109/LSENS.2020.2986794

[21]. Kennedy, J., & Eberhart, R. (1948). Particle swarm optimization. In IEEE International Conference Neural Networks. Perth, Australia: IEEE Press, 1995: 1942. DOI: 10.1109/ICNN.1995.488968

[22]. Shabir, S., & Singla, R. (2016). A comparative study of genetic algorithm and the particle swarm optimization. International Journal of Electrical Engineering, 9(2016), 215–223.

[23]. Gupta, S. H., Singh, R. K., & Sharan, S. N. (2016). An approach to implement PSO to optimize outage probability of coded cooperative communication with multiple relays. Alexandria Engineering Journal, 55(3), 2805–2810. DOI: 10.1016/j.aej.2016.07.018

Part II

Computational Intelligence for Urban Industry

5 Smart Parking Systems Using Image Processing

*Venkat Sai Karanam, Vinod Kumar Shukla,
Shaurya Gupta and Preetha V K*

CONTENTS

5.1 INTRODUCTION

Innovative screening services can be useful for public parking systems as they can cover many areas compared to an average ticketing employee. Smart screening systems use digital cameras, machine learning and a powerful artificial intelligent system to help screen and scan vehicles for violations regarding paid parking. This system uses the camera on top of patrol vehicles, ticketing agent vehicles and cameras surrounding the parking lot. There has been a massive increase in vehicles worldwide in recent years. Finding a parking spot/space can be difficult in crowded or busy areas. This leads to common conflicts like congestion, time delay and some accidents. In this chapter, we will be talking about and explaining a smart system that can be used to identify a free slot in a parking area and help with the payment for the given parking slot/space. This system can also help manage the parking lot/area to record the number of parked cars.

5.2 LITERATURE REVIEW

This chapter includes published research related to smart parking systems using image processing. image processing captures data of the parking lot and scans for space and occupied spaces. This will help the driver to find a parking spot quicker

DOI: 10.1201/9781003218715-7

than the traditional method. The automatic number-plate recognition (ANPR) system will help scan number plates and add the vehicle to the database of the parking area. The result of this research has helped with the understanding that smart parking will help save time and emit fewer carbon emissions for each vehicle that enters the parking area. It will also help with digitalization.

5.3   SMART PARKING SYSTEM OVERVIEW

The smart screening system scans each vehicle parked in the parking lot and checks with the paid parking database to check if the person has paid for the parking. This system also checks if the car is parked or if the vehicle is waiting at the parking spot. When finding a parking violation, the system sends the information to the nearest ticketing agent for re-checking the parking conditions if the vehicle is violating the parking rules. Further, upon verification, the penalty is sent to the fine issuing department (parking violations department) for it to be sent to the vehicle's owner by SMS or email. Digitalizing everything from ticketing to paying the violation has been digitalized as it reduces the wastage of paper and saves time for both the vehicle owner and the parking authority [1].

This system is currently being implemented in various places within the UAE. The smart detection system for public paid parking adds to the Dubai government's plan to digitize the customer services of public organizations in the emirate. It is also in line with the UAE's AIS (artificial intelligence strategy), which the UAE government supports. The system modernizes and regulates the use of public parking spaces and increases the frequency of the use of parking spaces. It also increases the efficiency of monitoring and handling of parking spaces as well as reduces possible errors in the event of violations. The execution of smart inspection of public parking spaces is part of a plan to facilitate the work of parking inspectors using cutting-edge technologies to monitor parking spaces' use [2–3].

5.4   AUTOMATIC NUMBER PLATE RECOGNITION SYSTEM

This system uses automatic number plate recognition (ANPR), which is a system that can scan number plates and update the vehicle owner on how much time or if there is payment due for the given parking lot [4].

ANPR technology is the best suited for open space smart parking solutions as it does not require using RFIDs or access cards. ANPR uses optical character recognition to scan for the number plate of each vehicle that enters the parking area/lot. It can use existing CCTVs, which reduces the cost of implementing this system. It captures the number plate and scans it for numbers and letters. Once it gets the number plate number, it checks if the driver has a membership for parking or is a prepaid user. The number then gets added to the database for payment of the parking. When the car parks into a parking spot, it gets charged hourly or daily, depending on how long the vehicle has been kept in the parking lot/area [5].

5.4.1   Sensors That Are to Be Used in Smart Parking Systems

Smart parking uses highly innovative hidden sensors (buried underneath or hidden above various parking spots) to monitor individual parking spaces and send occupancy status to the gateways, which in turn send this live status information to the online cloud platform for parking information which can be viewed in real time on multiple devices [6].

In this framework, sensors recognize the free parking spaces and parking slots, a vehicle is then recognized utilizing image processing, and the parking charge for each vehicle is identified based on the time that the vehicle is within the parking lot/area. It is helpful for the vehicles/drivers to find a free parking slot/space instead of wasting their precious time looking for a parking space/spot. This system will help drivers to reach their destination faster and help them to calculate the parking charge for the correct amount of duration for which the driver has parked his/her vehicle in the given parking spot/space. Payment for the parking spot/space is made easier as it is connected to the user's registered vehicle wallet, which can be recharged online or through a kiosk in the parking lot/area [7].

Smart parking uses various sensors, from cameras to sonar technologies, to detect vehicles and check for valid parking tickets. The vehicle detection system has the option of being drilled into the ground, and there is enough space for cabling. This might only be an option when the parking lot is being constructed, as it can be expensive to drill and set up the parking sensors. This system has an alternative option that can be used to mount sensors, such as surface mount sensors for closed locations in which it has surfaces that are over the vehicle, such as thin surfaces on cables, ceilings, and overhead lines. This system is used where there is less space available, or it is impossible to drill the ground [8].

These sensors work exactly like underground sensors and can be color-coded for better visibility in areas with heavy vehicle traffic. Another alternative is a more cost-effective, simple, and highly efficient intelligent parking system that can be used in multi-story parking buildings and does not involve any human interaction. It transforms multi-story parking into an automated parking system where the sensors are color-coded for drivers to find parking and help with parking payment services easily. At the entrance of the building, it scans for the number plate and registers it to the database, which the smart parking system can then use to scan for payment violations. This system is reinforced by cameras and various sensors [9].

5.4.2   Use of IoTs in the Smart Parking System

Connectivity of the system is used by IoT systems where multiple gateways units use operators. Operators have a compatible device connecting to other services like lighting, public Wi-Fi, surveillance, and more. It also means that less hardware will clutter city streets, and there will be a simplified way to manage a range of smart city services. The central server receives all the data logged in the input stage via a wireless transmission network. It ensures absolute protection during data storage activities, and user information cannot be traced as it is secured [10].

Some of the common uses of IOTs used in this system are:

- To scan vehicles
- To check for free slots
- To check if the vehicle has paid their fees or not.

Different kinds of sensors are in place within the given smart parking system/framework to help play a significant part in the smart parking framework. With the availability of various kinds of sensors for use, various kinds of characteristics are to be considered before being implemented into the smart parking framework. Some of the characteristics to be considered are [11]:

- Low cost
- The size needs to be flexible for different kinds of parking areas/lots
- Reliability
- Robustness
- Low maintenance
- High efficiency.

Various kinds of sensors can be categorized into two categories: invasive and non-invasive. Invasive sensors occupy huge space and require much manpower, increasing the cost of installation/maintenance. Non-invasive sensors are small and easier to install; they are cheap and easily available. These non-invasive sensors can be easily used in any parking lot as they can be fixed/mounted on the ground or the ceiling of the given parking lot/area [12].

5.4.3   WHY ANPR/IMAGE PROCESSING IS BETTER AS COMPARED TO PHYSICAL SENSORS

Image processing has been chosen as it is the best system with various advantages over invasive sensors. Image processing can easily detect a vehicle using any camera or CCTV installed in the parking lot/area. It can analyze and detect the change between frames. It can monitor wide spaces, multiple paths, and multiple parking lot areas and has lower maintenance. It is the best cost-effective method that can easily be installed and modified according to the user's requirements. It can process and capture anything within the camera's field of view. Other kinds of sensors can be used with the image processing system, such as microwave sensors and passive infrared, to help with a two-step type verification to check for vehicles in the parking area/lot [13].

All the devices are connected to the parking system's network, where all the information of the vehicles is stored. When a vehicle enters the parking system, the camera captures the vehicle's number plate and registers it to the system. Digital signage is available on each parking row to help assist the driver towards a free/empty parking slot. The user can either follow the digital signage or download the parking app to help find the exact empty spot in the parking area/lot. Once the vehicle is parked, the cameras capture and time stamp the vehicle. Once the time stamp is done, the car will be charged for the parking spot. After the vehicle is taken out of the parking space, there will be another time stamp so the vehicle can stop being charged for parking [14].

The image processing works uniquely, making it easier to run. When a car enters a parking lot, the smart parking system finds space and guides the driver to the free spot. The user can either use digital signage or the mobile app linked to the smart parking system.

5.4.4 WORKING OF THE SMART PARKING SYSTEM USING IMAGE PROCESSING

The software uses rectangles to define parking spaces; whether there is a car or not, the rectangle's color will change accordingly. Once the user parks their vehicle in a free slot, the camera scans if there is a vehicle or not, and then it changes the color of the parking spot to red if it is occupied on the smart parking system server. This system will be updated using cloud computing, and it will also be updated on the database for the vehicle parked. It adds the time stamp to the time the vehicle parked at the free parking space [15–16].

The software takes an input image from the given cameras and pins the empty or occupied parking spaces. To check if the parking space is occupied, it detects vehicles and then tells the system that the parking spot is occupied. Ultimately, it detects empty parking spaces and tells the system where they are located. Once this information is sent to the system, the system sends the necessary information to the digital signage and the mobile application. The management then uses this information to find out if someone hasn't paid for parking and occupied spots. For the driver or the user, the system shows the necessary directions to the empty slot on the app and, through digital signage, the number of empty parking spaces in the given parking space [17–18].

5.5 TECHNICAL REQUIREMENTS

5.5.1 HARDWARE REQUIREMENTS [19]

- RAM: 500 megabytes (GB) for 32-bit or 1 GB for 64-bit
- Hard disk space: 8 GB for 32-bit OS or 12 GB for 64-bit OS.

5.5.2 SOFTWARE REQUIREMENTS [19]

- Windows 7, 8, 10 and up, macOS, and Linux
- OpenCV.

5.6 CONCLUSION

The smart parking system is currently being used in various places in UAE. This technology is efficient, time-saving and has zero carbon emissions. This system can be enhanced with better artificial technology available later. This technology will not only enhance the parking experience for the user but management of the parking lot as it will help with the easier management of the parking area. This system considers low cost, high efficiency, low installation cost and robustness.

A smart parking system will help with the digitalization of the world. Digitalization of the world help with reducing carbon emissions. Future technologies like car superchargers will make it easier for drivers to charge their vehicles as they park their cars in a parking lot. Smart parking systems, when it is applied in various places, will improve artificial intelligence as it learns every time it gets data. The collection of data is going to help with the improvement of the image processing of the parking area/lot. This technology will help increase the density of smart parking and reduce waiting time or even collisions that are caused in parking areas [20–21].

REFERENCES

[1]. Idris, M.Y., et al. (2009). Smart parking system using image processing techniques. *Journal of Information Technology*: 114–127.

[2]. Government, U.A.E. (2020, October). UAE strategy for artificial intelligence. *UAE Strategy for Artificial Intelligence—The Official Portal of the UAE Government*. Retrieved December 13, 2021, from https://u.ae/en/about-the-uae/strategies-initiatives-and-awards/federal-governments-strategies-and-plans/uae-strategy-for-artificial-intelligence

[3]. The Report, G. (2019, May 8). RTA smart parking: Finding a parking spot becomes easier. *Transport – Gulf News*. Retrieved December 13, 2021, from https://gulfnews.com/uae/transport/rta-smart-parking-finding-a-parking-spot-becomes-easier-1.63819671

[4]. Patel, C., Dipti, S., and Atul, P. (2013). Automatic number plate recognition system (ANPR): A survey. *International Journal of Computer Applications* 69.9.

[5]. Fraifer, M., and Mikael, F. (2016). Investigation of smart parking systems and their technologies. In *Thirty-Seventh International Conference on Information Systems. IoT Smart City Challenges Applications (ISCA 2016)*, Dublin, Ireland.

[6]. Kianpisheh, A., et al. (2012). Smart parking system (SPS) architecture using the ultrasonic detector. *International Journal of Software Engineering and Its Applications* 6.3: 55–58.

[7]. Al Maruf, M.A., Ahmed, S., Ahmed, M.T., Roy, A., and Nitu, Z.F. (2019, January). A proposed model of integrated smart parking solution for a city. In *2019 International Conference on Robotics, Electrical and Signal Processing Techniques (ICREST)* (pp. 340–345). IEEE, American International University-Bangladesh (AIUB), Dhaka, Bangladesh.

[8]. Unnikrishnan, A., Uday, M., and Kaushik, G. (2018, April). Intelligent and prognostic parking system. *IAETSD Journal for Advanced Research in Applied Sciences* 5.4: ISSN No: 2394-8442.

[9]. Ruili, J., et al. (2018). Smart parking system using image processing and artificial intelligence. In *2018 12th International Conference on Sensing Technology (ICST)*. IEEE, Ireland.

[10]. Abdulkader, O., Bamhdi, A.M., Thayananthan, V., Jambi, K., and Alrasheedi, M. (2018, February). A novel and secure smart parking management system (SPMS) based on integration of WSN, RFID, and IoT. In *2018 15th Learning and Technology Conference (L&T)* (pp. 102–106). IEEE, Saudi Arabia.

[11]. Paidi, V., et al. (2018). Smart parking sensors, technologies and applications for open parking lots: a review. *IET Intelligent Transport Systems* 12.8: 735–741.

[12]. Idris, M.I., et al. (2009). Car park system: A review of the smart parking system and its technology. *Information Technology Journal* 8.2: 101–113.

[13]. Dsouza, K.B., Mohammed, S., and Hussain, Y. (2017, April). Smart parking—An integrated solution for an urban setting. In *2017 2nd International Conference for Convergence in Technology (I2CT)* (pp. 174–177). IEEE, India.

[14]. Polycarpou, E., Lambrinos, L., and Protopapadakis, E. (2013, June). Smart parking solutions for urban areas. In *2013 IEEE 14th International Symposium on "A World of Wireless, Mobile and Multimedia Networks" (WoWMoM)* (pp. 1–6). IEEE, Madrid, Spain.

[15]. Kanteti, D., Srikar, D.V.S., and Ramesh, T.K. (2017, August). Intelligent smart parking algorithm. In *2017 International Conference on Smart Technologies for Smart Nation (SmartTechCon)* (pp. 1018–1022). IEEE, Bengaluru, India.

[16]. Rane, S., Dubey, A., and Parida, T. (2017, July). Design of IoT based intelligent parking system using image processing algorithms. In *2017 International Conference on Computing Methodologies and Communication (ICCMC)* (pp. 1049–1053). IEEE, Erode, India.

[17]. Sunmathi, S., Sandhya, M., Sumitha, M., and Kirthika, A. (2019, March). Smart car parking using image processing. In *2019 5th International Conference on Advanced Computing & Communication Systems (ICACCS)* (pp. 485–487). IEEE, Coimbatore, India.

[18]. Khanna, A., and Rishi, A., (2016). IoT based smart parking system. In *2016 International Conference on Internet of Things and Applications (IOTA)*. IEEE.

[19]. Kanteti, D., Srikar, D.V.S., and Ramesh, T.K. (2017, April). Smart parking system for commercial stretch in cities. In *2017 International Conference on Communication and Signal Processing (ICCSP)* (pp. 1285–1289). IEEE, Chennai, India.

[20]. Vakula, D., and Yeshwanth Krishna, K. (2017). Low cost smart parking system for smart cities. In *2017 International Conference on Intelligent Sustainable Systems (ICISS)*. IEEE.

[21]. Alam, M., et al. (2018). Real-time smart parking systems integration in distributed ITS for smart cities. *Journal of Advanced Transportation* 2018.

6 The Internet of Things and Radio Frequency Identification in the Sports Industry

Bakari Juma Bakari, Vinod Kumar Shukla, Preetha V K and Deepa Gupta

CONTENTS

6.1 INTRODUCTION

The IoT (Internet of Things) in the sports industry has been progressing rapidly. It has advanced to the stage where it can detect our physical health by analyzing and recording our daily activities and concluding with a daily/monthly average of our activities [1]. IoT sensors can be used to determine ventilation, electricity, water, heating, and other electronic systems that are used in spaces such as stadiums. However, this chapter will focus on the use of IoT on sports equipment such as fabrics, insoles, shoes, and devices that athletes can wear. Keeping aside all advantages, some downsides need also be addressed. It is vital to consider the possibilities of security breaches where personal information can be leaked by hacking into those devices. Every application requires a profile and access to certain personal data to be able to be efficient [2]. Different sensors have conversed, such as [3]:

- Temperature sensors (measure the amount of heat energy in an object that records temperature variations over time);
- Humidity sensors (measure the water vapor in the environment and record the quantity of other gases);
- Pressure sensors (measures pressure fluctuations in gases and liquids);

DOI: 10.1201/9781003218715-8

- Accelerometers (measures acceleration, object velocity);
- Gyroscope (measures the rate of velocity in relation of speed around an axis);
- Infrared sensors (measures characteristics of their environment like heat, blood flow and pressure);
- Optical sensors (measure breathing and heart rates).

From business points of view it can be stated in favor of their economic goals, such as:

- *Efficiency*: Establishments can use traffic flow organization, building routine and stadium security to improve game experiences;
- *Experience*: IoT devices can intrigue fans and enhance their game experiences, which could be very profitable;
- *Revenue*: It permits businesses to use IoT to benefit from the drive people get from new technological advancements;
- *Personalization*: This can allow athletes to connect more with their fans by using IoT's advancements to share personalized routines and activities [4].

The IoT serves the same purpose as conventional media such as television, radio and newspapers. The timeliness, ease and a growing number of other advantages have been demonstrated in public as a medium with the greatest development potential for transmitting athletic culture. The IoT is a fast-expanding network of interconnected items that can gather and share data in real time using embedded sensors. The IoT can link thermostats, automobiles, lighting, refrigerators, transport management [5–7], supply chain management [8–9], tourism and hospitality [10–11], agriculture management [12–13], healthcare [14–17] and other items [18].

Football players always wear pads to which several sensors may be attached to collect data. RFID (radio frequency identification) tags on player pads have been used to measure speed, acceleration, distance run and distance from other players in real time. This is ideal for coaches who need to monitor how quickly their players react in real-time situations. The sports industry is adapting the technology continuously, and trending equipment is getting associated with the games for multiple purposes, such as the assisted referee (VAR) to support human wisdom with the support of the electronic system to measure performance in an accurate sense and plan better strategies for the player and game [19].

6.2   PURPOSE AND MOTIVATION

The IoT is a solution for thousands of processes inbuilt into the devices and objects which are connected to the Internet. The data in these devices can roam worldwide and provide accessibility to the devices to be controlled, monitored, and programmed remotely (Figure 6.1). The IoT and Internet have covered the worldwide devices in the network and connected the objects to the intelligence of the Internet, which has transformed many industries. Sensor chips and the Internet turn objects

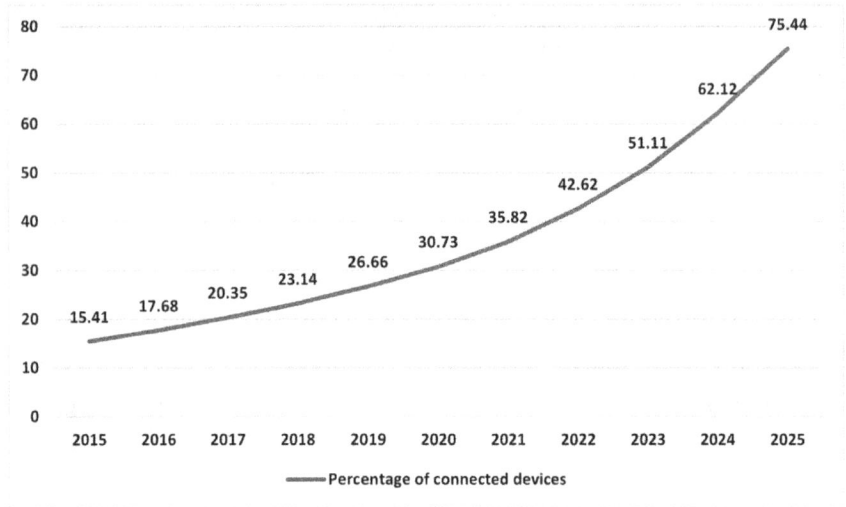

FIGURE 6.1   Upcoming demand of IoT devices (in billions) [6–7].

into intelligent forms and connect the physical to the digital world. The IoT is turning things more advanced and intelligent.

This has started communicating information from the device to the Internet and getting managed with its possible features. The smallest light bulbs to the jet aircraft engine are supported with the IoT technology for the sustainable and accurate prediction of the service requirements, utilization, replacements and regular operations through the IoT. Smart cities are recording, analyzing and understanding the overall environment with the help of IoT and aircrafts using multiple sensors on their wings to receive data transmit data and analyze the physical world with sensors [3].

6.3   CHALLENGES AND ISSUES OF TECHNOLOGY IN THE SPORTS INDUSTRY

Radio frequency identification (RFID) transmits data from RFID tags to an RFID reader using radio waves. A sensor is connected to an antenna on a sticker, transmitting data to the reader. A reader can scan more than 100 tags at once and not include a line-of-sight visibility. Each sensor usually contains unique identifiers. This simplifies procedures that would otherwise take more time and effort and can be vulnerable to human error. The system is efficient and multifunctional, durable and it can withstand weathering [20].

The disadvantages of RFID include signals getting jammed by metallic elements and fluids. This can be a huge problem for athletes as they sweat while training and playing sports, which could affect the system and record false data due to the sweat-jamming signals. Sometimes the information recorded is not as accurate as expected in other systems. Hence, the system has room for error. RFID are expensive systems that can be difficult to install and execute; thus, they are time-consuming [21–22].

Standards for RFID are currently being established. You do not want to buy an RFID system based on soon-to-be-outdated specifications. RFID technology eventually contains software that permits a central database to identify each user. Hackers will almost probably target this infrastructure. The tag may be damaged by water, static discharge, or high-powered magnetic surges (such as a lightning strike). Outdoor sports such as football and rugby can also have weather disturbances such as lightning and rain [23].

The shortcomings and limits of this technology can be eliminated with additional research. RFID technology will be extremely valuable in various industries, including retail and transportation. RFID technology advances, resulting in greater memory capacity, broader scanning ranges and faster processing. RFID will continue to expand in established segments where barcode or other optical technologies are ineffective. The market will increase enormously if standardization is reached, allowing RFID technology from different manufacturers to be used interchangeably. However, the connection between clothing and technology is not flawless. Some of the apparel is stiff, and some materials are not particularly breathable. This could be a problem as athletes must wear comfortable clothes that do not limit their movement to give them their most efficient speed while in a game.

6.4   RFID AND ITS CHALLENGES

RFID systems over the years have to seem multiple problems, and there have been methods to help fix those problems. Here are the most common problems faced and their solutions:

There can be interference among numerous readers when more than two readers are running simultaneously, and there can be delays and interference with the data coming through. Reader installation and debugging must meet the following characteristics to avoid interfering with one another: Distance between them needs to be more than 3 m, two nearby readers' operational frequencies should be adjusted at 920–925 MHz frequency hopping and reading time intervals should be staggered.

Short reading distance: Tthe most common distance range that the system can reach is 5–20 m. Sometimes the settings of the systems are off, which does not allow people to read data at the distance it should. To solve this:

• The reader's frequency settings need to be rechecked (frequency hopping should be selected as the working mode, with a frequency hopping range of 920–925 MHz)
• Ensure that the label and antenna polarization directions are in the same direction
• Ensure the label's surface is not coated with other materials, such as metal
• Examine the attributes of the RFID label
• Inspect the RF cable that connects the reader to the antenna.

Cannot read the card: The PC command cannot be transmitted to the reader if the serial cable or network cable is not correctly attached, or if the cable is not connected or the connection is not secure. To repair this:

- Inspect the RFID antenna SMA connection for tightness and damage to the label.
- Verify that the label follows the domain-related standard.
- Examine the label for any damage.

The RFID devices are sensitive to any liquid contact that could affect the data recorded from athletes. A possible solution to this could be altering the material that the devices are made of to be more weather resistant, so they can withstand water and electricity without interfering with the data transmission and collection. Materials such as organic photovoltaics with an elastomer coating that is stretchable and waterproof for washable applications can be coated on the devices to protect them.

6.5   RFID AND ADVANCE SENSORS IN THE SPORTS INDUSTRY

Data collection methods are framed for the major related areas, such as the use of RFID and advanced sensors in sports. From what point of view is the RFID or the sensor important, or how can this be more beneficial for the athletes or the sports industry?

6.5.1   RFID AND SPORTS

Players use RFID to reflect on and evaluate their performance. RFID trackers provide data that may be used to track an athlete's abilities and game performance. An NFL running back, for example, may be able to assess their injury by comparing their average speed before and after the injury. Radiofrequency identification is important because it allows computers to record information about what is happening in the real world and warn management when things aren't going as planned. RFID can be more efficient than traditional barcode technology by scanning several tags at once. Wireless scanners can rapidly recognize and collect data when within scanning range, and tags can hold more information per chip than a barcode. RFID sensors can be useful in any sport. RFID sensors have an advantage over GPS sensors used in sports training and performance because of their size. You don't want anything that might damage or damage the athlete, especially in a high-impact activity like football. Players may use RFID to reflect on and evaluate their performance. RFID trackers provide data that may be used to track an athlete's abilities and game performance.

6.5.2   ADVANCE IoT SENSORS IN SPORTS

Athletics clothes can include sensors that employ electromyography (EMG) to monitor muscle activity. Electrical impulses move outwards through the muscular body as the neurons in the muscle's core fire to tell it to contract. The degree of muscle engagement and reaction time may be assessed by monitoring the intensity and speed of these signals on the skin's surface. Sensors can attempt to make the experience of using EMG as natural as possible by incorporating the sensors into clothes. According to the different targets of individual athletics, the sensor technology can help athletes of all levels train better and smarter. By integrating muscle tension, sensors provide a complete picture of the athlete. Coaches can build better athletes using these sensors for training sessions, informing them throughout the training. This helps coaches to identify athletes who may be at risk of injury and help players improve their movement quality so they can perform better on the field.

6.6   CONCLUSION

IoT sensors in the sports industry impact it by enabling training, managing player performance, and addressing essential game features; the IoT has the potential to transform the way games are played. Coaches and management may combine the power of sophisticated analytics and game footage with the use of IoT in sports. IoT devices may help athletes enhance performance, track development, and assess efficiency. Teams use data from wearables and linked devices. They may enhance in-game strategy, analyze opponents' vulnerabilities better, and make better draft and trade decisions.

The future of IoT in the sports industry is vast. The IoT in sports will not be limited to assisting athletes, players, and coaches. The ramifications of IoT in sports will also be exploited to guarantee that the viewing experience is improved. IoT in sports is also expected to help fans, with certain ramifications already visible in the real world. In most stadiums, multiple LED screens of colossal dimensions are now a reality. The increased usage of IoT in sports will aid stadium upkeep. Sensors around stadiums can monitor the stadium's condition, allowing for preventive maintenance. IoT may also be utilized in parking lots to help fans quickly identify available places. Furthermore, IoT may be utilized to deliver traffic information before and following a game and allow stadiums to develop methods to better control traffic congestion. Sports worldwide are seeing a progressive decline in supporters who attend games. Improving the fan experience will ensure that sports continue to receive the adoration they need to thrive. This would necessitate a higher level of fan involvement at stadiums. IoT may be used to increase fan participation in sports, for example, by allowing them to put filters on their social media postings to emphasize the game they just watched. The primary goal of implementing IoT in sports for fan service is to guarantee that attending a game in a stadium outweighs other options for spending the day with family and friends. This would result in more revenue and aid in continuing the process. Television broadcasts are another significant source of money for sports. IoT can also help fans improve their small-screen experience. It will change the future for the better.

REFERENCES

[1]. T. Trayush, R. Bathla, S. Saini and V.K. Shukla, "Iot in healthcare: Challenges, benefits, applications, and opportunities," in 2021 International Conference on Advance Computing and Innovative Technologies in Engineering (ICACITE), 2021, pp. 107–111. IEEE, Noida, India.

[2]. L. Catarinucci, D. De Donno, L. Mainetti, L. Patrono, M. Stefanizzi and L. Tarricone, "An IoT-aware architecture to improve safety in sports environments," Journal of Communications Software and Systems, 2017, 13(2): 44–52, doi:10.24138/jcomss.v13i2.372.

[3]. D. Sehrawat and N.S. Gill, "Smart sensors: Analysis of different types of IoT sensors," in 2019 3rd International Conference on Trends in Electronics and Informatics (ICOEI), 2019, pp. 523–528. IEEE, Tirunelveli, India.

[4]. M.K., Gowda, A. Dhekne, S. Shen, R.R. Choudhury, L. Yang, S. Golwalkar and A. Essanian. Bringing IoT to sports analytics. NSDI, 2017, [Online Accessed on May 2023], https://www.usenix.org/system/files/conference/nsdi17/nsdi17-gowda.pdf

[5]. A.L. Madana and V.K. Shukla, "Conformity of accident detection using drones and vibration sensor," in 2020 8th International Conference on Reliability, Infocom Technologies and Optimization (Trends and Future Directions) (ICRITO), 2020, pp. 192–197. IEEE, Noida, India, doi:10.1109/ICRITO48877.2020.9197783.

[6]. A. Siraj and V. K. Shukla, "Framework for personalized car parking system using proximity sensor," in 2020 8th International Conference on Reliability, Infocom Technologies and Optimization (Trends and Future Directions) (ICRITO), 2020, pp. 198–202. IEEE, Noida, India, doi:10.1109/ICRITO48877.2020.9197853.

[7]. A.L. Madana, V. Kumar Shukla, R. Sharma and I. Nanda, "IoT enabled smart boarding pass for passenger tracking through Bluetooth low energy," in 2021 International Conference on Advance Computing and Innovative Technologies in Engineering (ICACITE), 2021, pp. 101–106, IEEE, Noida, India, doi:10.1109/ICACITE51222.2021.9404602.

[8]. L. Wanganoo and V.K. Shukla, "Real-time data monitoring in cold supply chain through NB-IoT," in 2020 11th International Conference on Computing, Communication and Networking Technologies (ICCCNT), 2020, pp. 1–6. IEEE, Kharagpur, India, doi:10.1109/ICCCNT49239.2020.9225360.

[9]. L. Wanganoo, V.K. Shukla and B.P. Panda, "NB-IoT powered last-mile delivery framework for cold supply chain," in T.P. Singh, R. Tomar, T. Choudhury, T. Perumal and H.F. Mahdi (eds), Data Driven Approach Towards Disruptive Technologies. Studies in Autonomic, Data-driven and Industrial Computing, 2021. Springer, Singapore, doi:10.1007/978-981-15-9873-9_22.

[10]. A. Verma, V.K. Shukla and R. Sharma, "Convergence of IOT in tourism industry: A pragmatic analysis," in Journal of Physics: Conference Series (Vol. 1714, No. 1), 2021, p. 012037. IOP Publishing, Goa, India.

[11]. W. Grobbelaar, A. Verma and V.K. Shukla, "Analyzing human robotic interaction in the food industry," in Journal of Physics: Conference Series (Vol. 1714, No. 1), 2021, p. 012032, IOP Publishing, Goa, India.

[12]. V.K. Shukla, A. Kohli and F.A. Shaikh, "IOT based growth monitoring on moringa oleifera through capacitive soil moisture sensor," in 2020 Seventh International Conference on Information Technology Trends (ITT), 2020, pp. 94–98. IEEE, Dubai, UAE, doi:10.1109/ITT51279.2020.9320884.

[13]. I. Nanda, C. Sahithi, M. Swath, S. Maloji and V.K. Shukla, "IIOT based smart crop protection and irrigation system," in 2020 Seventh International Conference on Information Technology Trends (ITT), 2020, pp. 118–125. IEEE, Dubai, UAE, doi:10.1109/ITT51279.2020.9320783.

[14]. L. Athota, V.K. Shukla, N. Pandey and A. Rana, "Chatbot for healthcare system using artificial intelligence," in 2020 8th International Conference on Reliability, Infocom Technologies and Optimization (Trends and Future Directions) (ICRITO), 2020, pp. 619–622. IEEE, Noida, India, doi:10.1109/ICRITO48877.2020.9197833.

[15]. T. Trayush, R. Bathla, S. Saini and V.K. Shukla, "IoT in healthcare: Challenges, benefits, applications, and opportunities," in 2021 International Conference on Advance Computing and Innovative Technologies in Engineering (ICACITE), 2021, pp. 107–111. IEEE, Noida, India, doi:10.1109/ICACITE51222.2021.9404583.

[16]. S. Vinod Kumar and A. Verma. "Model for user customization in wearable virtual reality devices with IOT for "low vision,"" in 2019 Amity International Conference on Artificial Intelligence (AICAI), 2019, p. 201. IEEE, Dubai, UAE.

[17]. S. Vinod Kumar and A. Verma. "Enhancing user navigation experience, object identification and surface depth detection for" low vision" with proposed electronic cane," in 2019 Advances in Science and Engineering Technology International Conferences (ASET), 2019. IEEE, Dubai, UAE.

[18]. A. Meola, A Look at Examples of IoT Devices and their Business Applications in 2021, [Online Accessed on May 2021], www.businessinsider.com/internet-of-things-devices-examples

[19]. TechnoGym, The Benefits of Using Technology in Sport, [Online Accessed on May 2021], www.technogym.com/gb/wellness/the-benefits-of-using-technology-in-sport/, Technogym

[20]. V.K. Shukla and R. Gupta, "Enhancing user experience for sustainable fashion through QR code and Geo-fencing," in 2019 International Conference on Automation, Computational and Technology Management (ICACTM), 2019, pp. 126–131. IEEE, Dubai, UAE, doi:10.1109/ICACTM.2019.8776704.

[21]. V.K. Shukla and B. Singh, "Conceptual framework of smart device for smart home management based on RFID and IoT," in 2019 Amity International Conference on Artificial Intelligence (AICAI), 2019, pp. 787–791. IEEE, Dubai, UAE, doi:10.1109/AICAI.2019.8701301.

[22]. P. Agarwal, V.K. Shukla, R. Gupta and S. Jhamb, "Attendance monitoring system through RFID, face detection and ethernet network: A conceptual framework for sustainable campus," in 2019 4th International Conference on Information Systems and Computer Networks (ISCON), 2019, pp. 321–325. IEEE, Mathura, India, doi:10.1109/ISCON47742.2019.9036209.

[23]. SI, Athos High-tech Biometric Clothing Offers Wearable Fitness Tracking [Online Accessed on May 2021], www.si.com/edge/2015/06/15/athos-smart-apparel-fitness-tracking-gear

7 Smart Innovation in the Hospitality and Tourism Industry

Amit Verma, Wonda Grobbelaar and Vinod Kumar Shukla

CONTENTS

DOI: 10.1201/9781003218715-9

7.1 INTRODUCTION: SMART INNOVATION IN HOSPITALITY AND TOURISM INDUSTRY

The tourism and hospitality industry is a complex amalgamation of multiple differently operated industry units clubbed together for the tourist guest or visitor experience, outlining the service industry composition. The industry made approximately 2.9 trillion USD of direct contribution to the global economy in 2019 [1]. The industry is ever evolving with the varied needs and demands of the end consumer or business requirements. The industry has witnessed technological growth at its best. The pace of the industry adaptation has made the innovation a living truth for passengers. The industry is a forerunner in adopting, implementing and accessing its boundaries for the cost-effective, convenient, experience-oriented approach for the consumer and the business. The industry has revolutionized its existence with the latest technologies like robotics, IoT, artificial intelligence, virtual reality and augmented reality. With the inception of such innovative implementations, the value chain, process and business operation have been widely altered from the traditional approach. However, apart from implementing technological advancement in smart cities, the transitional phase of technological implementation in developing smart cities and the limitation of adaptability of such innovations in underdeveloped cities across the world is one of the constraints in achieving maximum mileage of innovations.

Innovation is the primary indication of technology's possible features to be used mindfully to exploit the maximum benefits for the business operation. The technological implication of useful technologies derives from the possible utilization of multiple processes for the business. The implication requires the adaptability of all stakeholders to the scenarios. For example, the robotic receiving operation in a hotel facility will limit the suppliers to producing the product labels in the required settings or are fully automated service providers. Similarly, the collection of tourist data is helpful only if it may assist them in real time due to ever-changing situation-based requirements.

7.2 OVERVIEW OF TECHNOLOGICAL TRENDS IN THE TOURISM AND HOSPITALITY INDUSTRY

The industry has witnessed multiple trends and has reshaped its structure to outperform. Initially, the industry made its milestone moves through the invention of railways, aircraft and so on which reduced the idle time for passengers and staff. With the changing requirements and preferences of the guest, tourists or consumers evolved, which translated into the adaptability of the latest trends. However, technological advancement grew exponentially after the Internet became common. This has brought a few significant turns to the industry.

As depicted in Figure 7.1, the trends of the technological adaptation of the tourism and hospitality industry through the indication of a few technologies have dramatically impacted the industry. The Internet and mobile usage have led the social platforms for communities to discuss, review, plan their travel and book it online with human-assisted or computer-enabled booking support. This has started new business models like Airbnb and Expedia online travel agencies. The innovation has

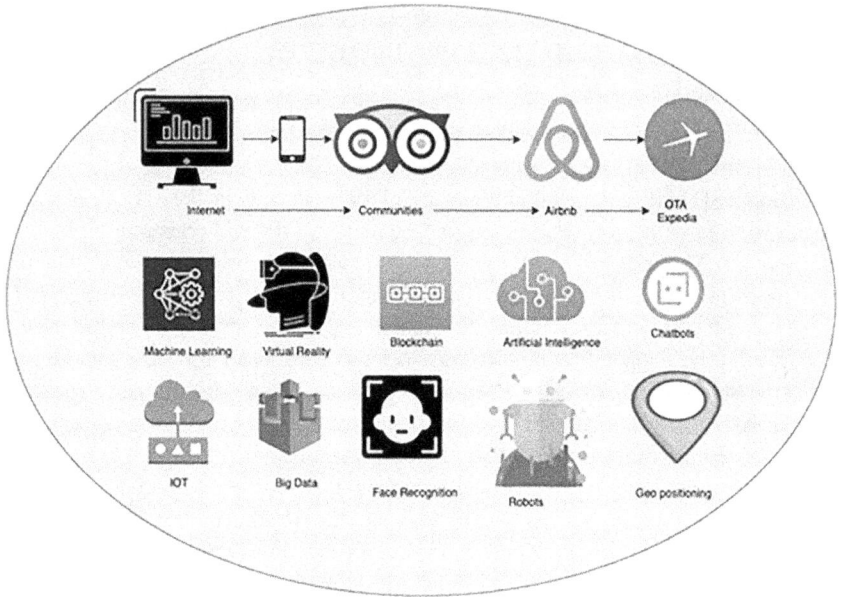

FIGURE 7.1 Technologies and tourism industry trend.

continued its expansion and turned the industry into a diverse platform for stakeholders and consumers with immersive technological advancements compelling to smart tourism requirements. The major technologies are artificial intelligence, machine learning, face recognition and analytics, robotics and blockchain.

The worldwide emergence of technological advancements and the inception of smart cities has opened a new highway of the possibility of growth not only for the non-smart cities but also for the industries of the smart cities. Industries have to transform to align themselves to the exceeding needs and requirements of the tech-savvy internal and external customers. The b2b or b2c business requirements have been changing the operation and shifting its technological framework to reach closer to the customer.

AI has impacted navigation apps, home and smartphone personal assistants, predictive maintenance and automation etc. Robotic process automation and machine learning are subsets of artificial intelligence, natural language processing and deep learning [2]. Information communication technologies (ICT) have been contributing to food security. The latest development of robots fortified with smart vision and soft grippers can multitask, understand verbal commands and add value to operations. The cost of robots has decreased significantly over the past few years and is fairly easy to install in any application. Robotics plays a vital role in process improvement, creating leaner operations throughout the hospitality and tourism sector. Therefore, integrating robots raises the bar on quality standards and solves many issues contributing to higher costs and less net profit. As proven in many different sectors, robots can assist with many tedious issues, including staff shortages, hygiene and

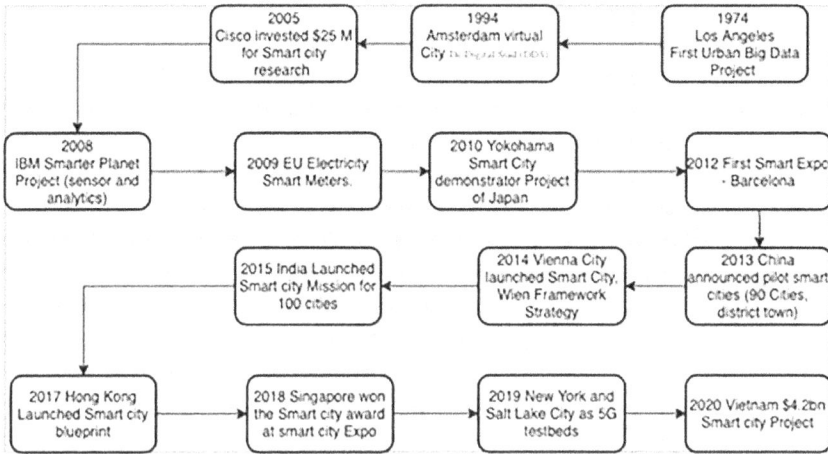

FIGURE 7.2   Timeline growth of smart cities.

ergonomic issues. Rapid progress in the development of robotics is creating numerous possibilities for robotics in the hospitality and tourism industry.

7.3   SMART CITY

Applications of Internet-enabled ICT (information and communication technologies) have provided a secure and smooth platform to operate in the smart city. A smart city can be defined where the effective usage of ICT is used to connect physical, social, business and governmental infrastructure to collect, transmit and utilize data as useful information for the convenience, safety, security efficiency and quality of overall life and to sustain the environment, social and economic aspects of the city [3].

The growth trend of the smart city (Figure 7.2) as discussed in the smart cities thematic research report by Global Data Thematic research is comprehensive to analyze the growth trend over the years [4].

7.3.1   SCOPE, CHARACTERISTICS AND ADVANTAGES OF SMART CITY

The components of the smart city are vast. They can be summarized with the indicative depiction of the city's essential components such as individuals, the public sector, transportation, airports, airlines, buildings, infrastructure, banking, services and education. The scope of any smart city is limited to its society, economy, environment and governance. The smart city provides a smooth, convenient and informed operation to both operator and provider to enhance the excellence day by day to minimize the wastage of resources and energy in the home business and public infrastructure. The benefits of the smart city are sustainability, quality of life, urbanization and smartness.

In a smart city, municipalities and government practice ICT to upsurge functioning competence, minimize variation, distribute communication with the community and advance the character of authority-related services and inhabitant benefit in multiple ways. The United Nations aims to transform the world by 2030 through its 17 Sustainable Development Goals (SDGs) and to set the trend for cities worldwide to become more sustainable; many examples could be identified. A good example of such a city is Bristol in the UK. Bristol focuses on five themes: energy, food, nature, resources and transport. The main focus of Bristol is to make it a healthier and happier city, including a low or no carbon emission city.

It was named in 2020 as the world's vegan capital for the third consecutive year. Bristol is committed to clean transport and energy and plays an important role as a low-carbon hub of industry. Bristol influenced international policy during the UN Climate Change Conference in Paris in 2015 by sharing best practices and presenting impressive sustainable accomplishments. Some included tackling food waste into delicious, ethical food on the streets in different activities.

The Bristol Food Network plays an important role in informing and encouraging locals to transform the city into a sustainable city. Municipalities transform open spaces to make Bristol an urban nature reserve.

The environmentally friendly cities are designed to avoid all negative impacts on the earth, natural resources and environmental surrounding, which is a responsibility of every generation to represent as they have received to the next generation. This also advocates the non-destructive "green living" lifestyle to constantly think and plan for nature and the environment.

Amsterdam is one of the cities where cycling is the main form of transport. A visit to the train station of Amsterdam Central will show thousands of bicycles securely parked for the day while people are working. The pollution is less and the air much cleaner than in other parts of the world. Several cities have adopted the usage of electric vehicles to reduce carbon emissions through electric vehicles. However, Amsterdam has further planned toward it and has adopted renewable energy with its trains, buses and subways. Inhabitants are spinning toward solar panels as an alternative to profitable renewable energy. The city encourages construction companies to build greener properties to secure land tenders.

Amsterdam has been listed as the Green City of Europe with an outstanding vision and adaptation of policies. Cycling is the main transportation medium. More than 75% of the population owns a bicycle. The studies show that the residents preferred bicycles over their cars (average of 0.87 times a day and their car 0.84 times during 2005–2007) (European Union, 2020). Amsterdam is also a sustainable city where you can find many parks, gardens and canals (Figure 7.3).

7.3.2 SMART CITY INFRASTRUCTURE

The infrastructure of a smart city will depend majorly on three components. City infrastructure, business or office infrastructure, and public infrastructure will constitute the smart city infrastructure and lead to a smart urbanized city, improved quality of life and sustainable environment. The connectivity of all these three infrastructures will gather the relevant information, which can be analyzed to organize and predict the possible preferred automatized and convenient solutions for daily life.

FIGURE 7.3   Example of a sustainable city: Amsterdam.

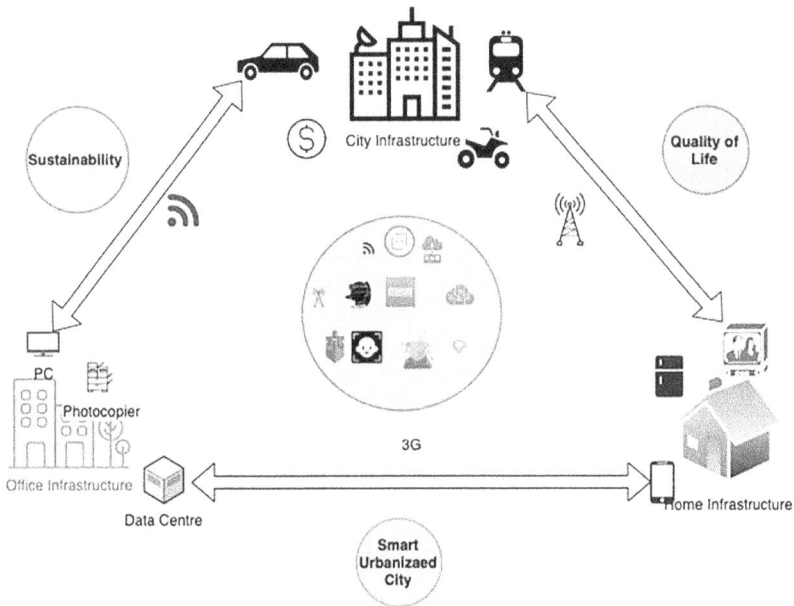

FIGURE 7.4   Smart city infrastructure.

The infrastructure holds the physical presence of the above-stated detailed spaces (Figure 7.4). At the same time, it is a combination of physical spaces connected through ICT for the services in the city. Innovative technologies are the backbone of the smart city and are fundamental to smart cities' design, implementation and operation. It provides a linked societal, economic, environmental and governmental collaboration for sustainable goals.

7.3.3 SMART CITY TECHNOLOGY

The basis of smart cities is the smart technologies and their implementation. Various technologies are available, and at the same time, various components are included in the design and operation of the smart city such as building, transportation, electronics, various analog and digital communication devices, and their software and compatibility. This has been installed in smart cities for regular updates and maintenance throughout the railway, flight and public transportation industries.

7.4 INNOVATION IN TOURISM AND HOSPITALITY INDUSTRY

In the 21st century, the hospitality and tourism industry has seen a transformational development through technological innovation and its smart combinations of technologies to optimize resource utilization and enhance efficiency after being environmentally responsible. One of many is facial recognition which has reduced the waiting time of passengers at immigration and reduced the collective loss of time for all. Another substantial combination of artificial intelligence and the Internet of things has reduced the unnecessary repetitive work in multiple areas of operation. The hotels have provided the luxury of keyless entry to the guests, and a few have transformed their full operation into robotic implementation. Moreover, technology like blockchain has been impressive in securing data from IoT and other connected devices [5].

7.4.1 IMPORTANCE OF INNOVATION IN THE INDUSTRY

The definition of innovation is important to be understood in real context to analyze the importance of innovation in the industry meaningfully. Innovation is a novel approach to creating values through a combination of solutions for significant problems [6]. The types of innovation can be explained through the two vertices of problem and domain, resulting in the following.

Figure 7.5 illustrates four styles of innovation: groundbreaking, primary research, breakthrough and disruptive research. However, sustaining innovations greatly support improving the quality of life [7]. Furthermore, the resulting face of industry due to these studies shapes an economy of refined operations and convenient service.

7.4.2 INTERNET OF THINGS: A GROUNDBREAKING TECHNOLOGICAL INNOVATION

Internet of things, commonly called IoT, is the networking capability that allows information to be sent and received from objectives and devices using the Internet. It refers to a system of interrelated, Internet-connected objects that collect and transfer data over a wireless network without human intervention. The personal or business possibilities are endless and forever changing the business world. The IoT is important as it creates huge savings for companies by automating processes and reducing the cost of labor. It also helps to reduce wastage and improves service delivery, making it less expensive to manufacture and deliver goods and providing transparency into customer transactions. There are six leading types of IoT:

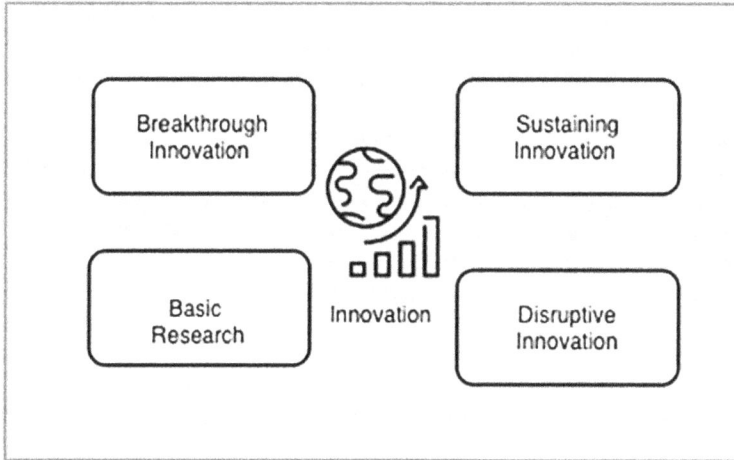

FIGURE 7.5 Types of innovation.

- LPWANs (low power wide area networks)
- Cellular (3G, 4G and 5 G)
- Zigbee and mesh protocols
- Bluetooth and BLE
- Wi-Fi
- RFID.

7.4.3 IoT FRAMEWORK AND APPLICATIONS

Several industry businesses and household areas have actively utilized the IoT. The major projects are announced in manufacturing or industrial setup in transportation, energy, smart home supply chain, healthcare and services. The platform types broadly utilized are application enabled, device management and advanced analytics with processors, semiconductors communication hardware, asset connectivity and cloud storage. The industrial IoT include industrial automation, smart factory solutions using wearables, production monitoring usage of augmented reality for the shop floor, unscheduled delays avoidance, automated or remote control of quality and other control systems (Figure 7.6). Tesla is one of the examples of connected transportation acting as a benchmark for many other car manufacturers like Volvo and Honda.

Global energy consumption is estimated to grow by 14%; hence, smarter energy solutions are always needed. Already smart grid optimization is in operation, and smart meters use machine learning algorithms using sensors. For example, Exelon, an American utility company, has utilized GE's Predix platform to improve wind farm accuracy to 70%. Similarly, IoT-enabled applications can easily monitor inventory management, smart vending kiosks and customer tracking [8]. Likewise, healthcare and smart city agriculture are equally involved in IoT growth.

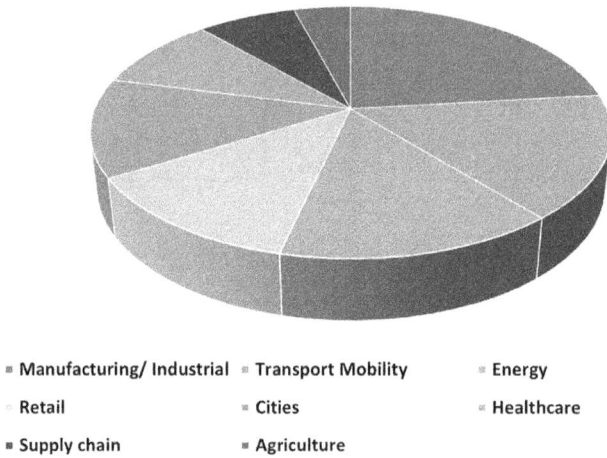

Manufacturing/ Industrial   Transport Mobility   Energy

Retail   Cities   Healthcare

Supply chain   Agriculture

FIGURE 7.6 IoT application area 2020.

Source: IoT Analytic Research [8].

7.4.4 INTERNET OF AIRLINES

EasyJet is a British budget airline specializing in affordable travel. It operates more than 600 routes across more than 30 European countries. EasyJet is always open to new ideas and innovation. One such project is wearable technology. EasyJet has been transformed by fitting staff suits with LEDs on the shoulder and edges to provide optical guidance to passengers with attached microphones for uninterrupted or undistorted communication with passengers and crew members. With the help of LED, the uniform displays the basic information about the flight, destinations and navigation support in case of emergencies. In another project, EasyJet uses drones to inspect its fleet of airplanes. Passengers can check in for their flights and download boarding passes, including a check for flights with an easy-to-use app. The app will also assist in finding out if there is an earlier flight back home. The change can be done from the customer's mobile app for a small fee. Once the boarding pass is downloaded, the boarding passes can be stored offline; therefore, no network is required. Sensors strategically placed in aircraft can be used to track anxiety amongst passengers.

7.4.5 INTERNET OF TOURISM

The Internet of Things can facilitate mechanization and a greater individual experience in travel and tourism. It can create a "lean culture" and streamline day-to-day activities in a hotel's daily running. The use of smart devices can reduce non-productive work for employees and can reduce energy costs. Alexa is a voice assistant that is linked to property-management systems. Guests can control lights and change TV programs, and Alexa can also answer questions about the property.

It enhances the customer experience and changes tIence for the customer forever. Similar technology can also be used in air travel, for example, adjusting seat temperature or controlling air conditioning. Creating a lean culture in all aspects of tourism and IoT engrosses restructuring much of the customer experience. From the first entry point to the airport, checking in at the counter, the experience while waiting for the departure, till the arrival and going through customs—basically throughout all industry ranges.

Basic sensors can be used to send communication to passengers' devices to alert them of the status of the baggage location and provide an easier and faster way to detect them. The check-in process at hotels can be streamlined; less time can be spent completing forms, and room keys can be shared on the smartphone to minimize the hassle of standing and getting details from the hotel counter. Moreover, the usage of the sensors may also keep restaurant staff attentive and informed with the arrival information and related food preferences, allergens, guest expectations, departure dates and timings and so on to enable businesses financial benefits and cost-efficiency.

Room temperatures can be kept or adjusted continually. In this way, heating is only used when it is needed. It is already scientifically proven that the most efficient room temperature is 24°C [9]. Similarly, the intensity of natural light in the room and movement detection may reduce consumption and, in turn, use less energy. IoT is also an effective tool to share information related to the location and acts as an important source of valuable data. This combination of smartphone capabilities based on location may lead to the appropriate coverage of relevant information to the suitable set of tourists or individuals at the real. Information about local attractions and times and special offers at certain times or at least busy can be sent to tourists. Some information can also include directions in the form of a map to simplify it for the tourist to reach the attraction. Information about guests using certain facilities in hotels can also be collected to improve the experience for guests in the future, change the offering to the guests if this is not popular, or have the correct staffing levels at any point in time.

The IoT can also be used for maintenance and repairs by delivering precious, real-time communication about their up-to-date status and equipment functioning in rooms, kitchens, and operations. Preventive maintenance provides pre-calculated information about the equipment that needs replacement or repair so as not to disrupt the operation with a fatal breakdown.

The technology implementation is exceptionally fruitful to the high expenditure areas of the industry such as airlines wherefrom the accurate fueling, engine parts usage and maintenance related information, and safety of passengers. Ultimately, the IoT has connected appliances and devices and enriched the ability to communicate, leading to superior customer focus and restructuring processes to offer numerous benefits.

7.4.6 INTERNET OF HOTELS

IoT in the hotel and tourism industry aims to create a seamless experience for the guest to attract more business and increase cost savings. Guests will feel more comfortable and remember their experience forever due to the excellent coordination of choices and individual requests based on their previous stay with the hotel chain or the hotel itself. The integration of guests' devices and the profile building of every individual guest support customization of reservations, easily implementing the stay preferences and choices of the repeat customer and prompts easier access to the possible customer's requirements through relevant feedback and surveys in a real-time environment. The information would support the hotel managers in planning the event and managing the different requirements to ensure the comfort and safety of the customers. An example can be the alteration of temperature based on the number of attendees in a ballroom, or the change of the air with the purified air in the room after a certain duration would reduce the chances of contamination.

7.4.7 INTERNET OF KITCHEN

The IoT of kitchens has transformed the devices through the intelligent usage of artificial intelligence (AI) sophisticated machine learning algorithms; combined with interfaces, the interactions are smoother and more accessible in real time. The technologies have made it more convenient, less expensive, or more effortless; however, they are reshaping the fundamental structures. The IoT has brought smart kitchen gadgets, using intelligent sensors that remind us to slow down the stove to ensure optimum utilization. The online ordering of groceries was unheard of a few decades ago, but today, barcodes and devices let consumers use them to manage shared grocery lists. In addition to scanning barcodes, some gadgets can accept voice commands and even place grocery delivery orders online. One such example is Amazon Dash; however, it is not freely available in all areas. Another example is Hiku, similar to a large fridge magnet. This device integrates with several suppliers. In the future, this device can make online price comparisons and orders. Some devices help to keep an eye on things and can pass alerts about a problem, for example, excessive smoke, traces of carbon monoxide, excessive temperatures and humidity levels. Some devices check for gas buildup from a leaky or forgotten stove. This information can be sent to any user, anywhere in the networked world, including local emergency departments. Humans are visual creatures. Virtual reality can create storylines to enhance the guest's dining experience through storytelling. Augmented reality can display models of food in a 3D form to create a realistic experience and, in this way, increase profits. The guest also has a better understanding of food that might be unfamiliar to him.

Smart dustbins are used in kitchens to control food waste and generate a more detailed comprehension of wastage. This digital solution gives chefs information on what is going on in their dustbins and what food could still be used. Software analysis of the waste of each discarded load daily and give regular reports on financial cost and sustainability. Other devices on the market are cups with internal sensors that analyze ingredients to show how much and how often you drink. The kitchen jar is also changing. These days kitchen jars can measure dry goods and how much is left with a built-in digital scale. It can tell you when it is running out of stock, give suggestions on recipes, and calculate the nutritional value. It will also place orders online when the minimum acceptable level is reached inside the jar. In another example, an egg minder is a device that will inform the chef how many eggs are left on the egg tray and which eggs are the freshest (E&T editorial staff, 2020) [10].

For an inventory, the question remains: how much raw material is available and how much product is needed? IoT opens a new way of managing inventory in real time. Communication is improved, and 5G has even enabled micro-tracking. The best part is that very little human interaction is required to create seamless dry store management. Algorithms can track and manage inventory better than their human counterparts. IoT creates an easier pathway to integrate robotics in future. Voice-activated assistants are limited to the kitchen and can assist in many ways. This can become your ideal sous chef. These voice-activated assistants can quickly answer difficult questions, scale recipes, calculate difficult calculations, and even explain an unknown culinary term if you cannot find the answer.

Smart devices bring Bluetooth-connected technology into the kitchen so that appliances can communicate. An appliance is smart if it enables user preferences or external signals remotely through intelligence and communications utility. Some examples include refrigerators and sous vide cooking devices, where the food is simmered in vacuum-packed plastic bags with the water steam once it is ready to be cooked. It will hold the food at a temperature of refrigeration which may be adjusted with the mobile app and timers to serve it at the appropriate time and temperature.

Refrigerators can be monitored outside the kitchen to ensure the temperature stays consistent. The chef can control the setting and the outcome with an app to deliver the best-tasting and consistent results. Food thermometers can be used in the oven or grill. Once the temperature reading is taken, it can be stored on a computer in the form of a report. For some devices, the chef only needs to decide what recipe to make today, and the rest will be automated. One of the advantages of an IoT kitchen is that if in the future, enough basic functions in the kitchen can be automated, chefs can take on more advanced roles and improve the skills required for their future job. Critical thinking and creativity will become more important to be successful in back-of-house operations.

7.4.8　Internet of Attraction (Using IoT and Big Data for Tourist Attractions)

The Internet has a significant impact on tourism for both providers and consumers. One such example is the Sagrada Família in Barcelona [11]. The city is renowned for presenting the Mobile World Congress (MWC) ("IoT & Big Data—Sagrada Familia, d-LAB—MWC," 2020). The city must live up to this important event, and it is no surprise that it has been formulating IoT and big data solutions to improve its city.

The logical implementation of relevant technologies has improved the tourist experience and movement at the Basílica de la Sagrada Família, the famous UNESCO World Heritage Site in Barcelona, Catalonia, Spain [11].

All the stakeholders involved reckoned the need to understand the flow of tourists in the attraction and nearby area. It has resulted in the further analysis of the tourist motive flow liking and disliking etc. This smart city project was an initiative of the Mobile World Capital's d-LAB digital. The project was conducted during the annual MWC fair, and the trial served as a way to support and show off mobile-centric innovation in Barcelona to the conference attendees. It also enabled d-Lab to better understand data on the number of visitors to this tourist attraction, such as the most popular time of day to visit and the age groups. It also provided insight into how the presence of the tourists impacted the local environment and what type of transport they used to reach the site. The Report of the collected data was analyzed to create the best way forward to preserve these attractions. Firstly d-LAB was set out to collect as much information as possible from tourists visiting this attraction. They aimed to be to create a visitor profile.

There are different profiles, such as tourists, travelers, day visitors, or people who visit Barcelona only for the nightlife. Applied data from the mobile group Orange was used to analyze the tourist's movements. The most commonly followed routes taken by tourists were also recorded. Some of the routes included the Ciutat Vella and Sants-Montjuïc districts. After collecting the data, the IoT elements were implemented to evaluate and determine street-level mobility patterns. This data was collected over 4 weeks. Wi-Fi and 3D sensors were located in this area. These sensors identified the main entry and exit points of the site. The maximum pedestrian crossing was located at Carrer de Mallorca and Carrer de Marina. It was found that 50% of that who went to the Sagrada Família stayed in the area for less than 40 minutes. Only 20% enter the church itself. Through the collection of this data, the busiest times were also revealed. The mornings from 10:00 am till midday were the busiest. In the afternoons, there was a spike in visitors to this site. After this project, recommendations could be made to the city manager to improve the tourist's experience and utilize this site more efficiently to generate income and protect this world heritage site.

7.4.9 INTERNET OF DESTINATION

The destination is a product for the tourist visiting served in several forms and services throughout the stay. The arrival to the smart airport handling smart check-out procedure and then the smart car booking service and smart drive to the smart hotel is not only the end of the smart destination. However, the destination aspired to significantly utilize each guest's data in detail to analyze the overall consumer behavior and destination planning to impact the GDP through multiple repeat visits of satisfied tourists.

In this context of the smart destination approach (Figure 7.7), each touchpoint of the tourist becomes a mine of data and preferences, leading to the smartly managed destination. From the choices of attraction to the waiting time, expenditure, overall experience, retail and service operations are to be altered with the right analysis of each choice [12].

7.5 ROBOTICS IN TOURISM AND HOSPITALITY

Robots are not the future anymore. They are a living reality in the world. The significant growth of robotics in industrial operation and associated benefits to the manufacturer and customers made it a futuristic plan. A robot is a programmable machine with its multifunctional sets of actions and manipulations designed to move, assemble or maneuver the material, parts, tools or specialized devices through multilayered programs inbuilt for tasks, motion or action.

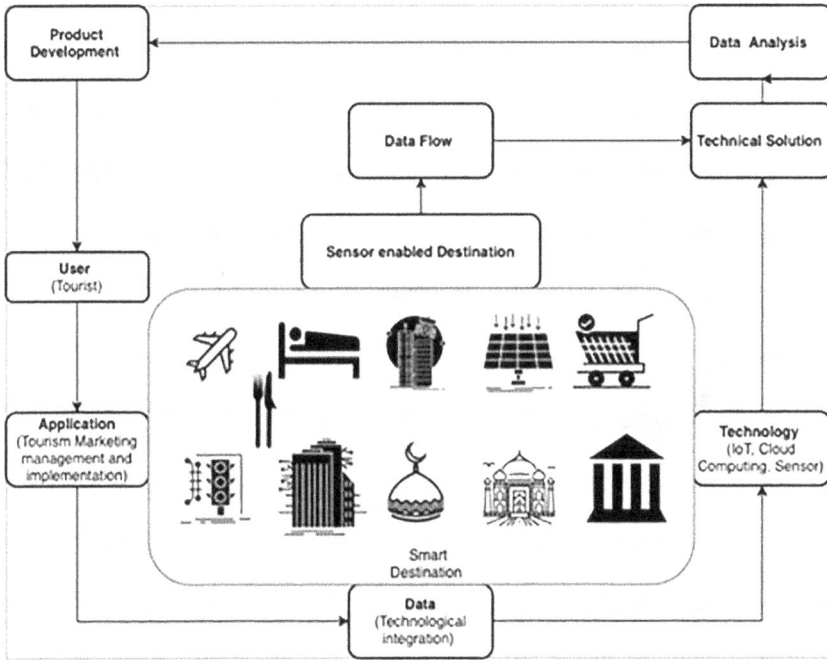

FIGURE 7.7   Internet of destination.

Source: Analyzing influence of IoT in tourism industry [12].

The definition provides a broad aspect of the robot. Clarity can be brought in by understanding the difference between automatic and robotic. Any automatic machine can be programmed by computer or manually but still will not be termed a robot unless it adjusts its action as per the outer signals received by the sensors and relative actions are taken.

7.5.1 ROBOTS AND ITS TYPE

The robots are categorized into three categories: mobile, manipulator and hybrid. The mobile robots can be differentiated by wheeled, leg, underwater, autonomous or aerial. Industrial robots are classified into six major types based on their axis movements [13]. The specific ones are depicted in Figure 7.8.

1. *Articulated*: It is a robot with commonly six to seven axis articulations used for picking and placing, assembly, welding, packaging, palletizing, inspection, material removal and dispensing. This has a high-flexibility mountability on all surfaces and easily operates between non-parallel surfaces or obstructed areas to maximize its reach.
2. *SCARA*: Selective compliance articulated robot arm operation is fixed and can move in two axes only and is usually used for inspection, packaging and dispensing.

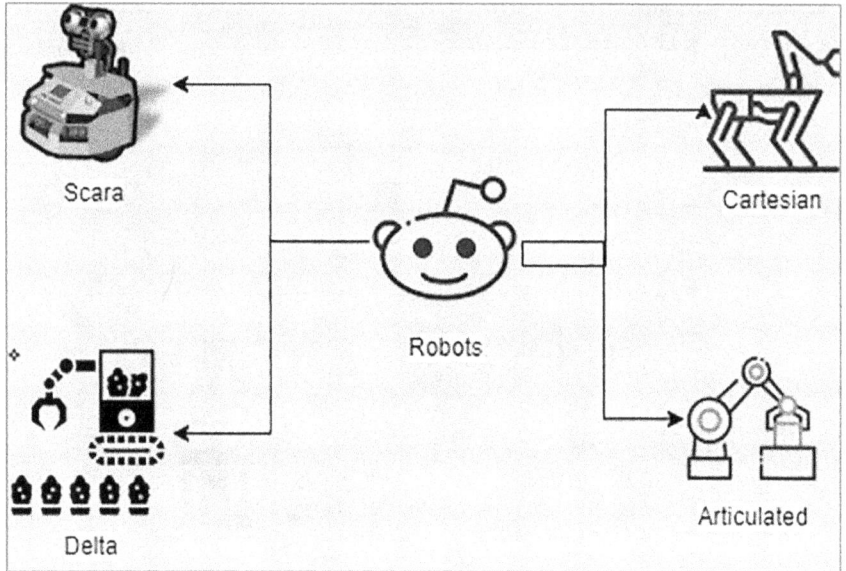

FIGURE 7.8   Main types of robots.

3. *Delta*: Delta robots or spider robots move in three axes in a circular diameter of operating range. These robots are very well implemented and used in the assembly line over a conveyor.
4. *Cartesian*: The cartesian robots, also known as gantry robots suspended over two parallel rails, cover the entire production floor in crane style structure. This is relatively low cost and covers large areas of work.

Moreover, the distinction that can be classified into collaborative and industrial robots must be understood. The cobots (Figure 7.9), or collaborative robots, work at a comparable speed to humans. They are flexible for redeploying several applications with high accuracy and the lowest implementation cost with in-house setup and programming capability. However, they do not have industrial robots' speed which seems not a mandate in non-industrial settings [14].

This may lead to several other combinations and configurations of the robots [13]. The robots are used mainly to avoid tedious, stressful, menial and labor-intensive, repetitive tasks or risk-prone jobs like working in contaminated areas and can be developed in several ways depending upon the requirements.

7.5.2   TECHNOLOGICAL APPROACH

Robots can be further subcategorized based on their technological approach to power, safety speed and hand guiding. As examples, of the power the collaborative robots in an industrial environment, they can be used to avoid collisions using joint sensors (used to monitor the forces applied), skin sensors (to watch out for its body

FIGURE 7.9 Cobots.

Source: Universal Robots and Cobots Basics; elixirphil.com [15].

and instructs to stop with the help of tactile sensing) and force sensor bases (to manage the torque and work among humans), and are inherently safe (handle low payload and cannot technically injure) [16].

7.5.3 MERITS AND DEMERITS OF ROBOTS

Robots' advantages are consistency, cost efficiency, speed accuracy and no fatigue. However, the actual usage or the advantages are seemingly higher than this. Robots can be used to do a task where human reach is limited or may lead to loss of life and any dangerous environment such as radioactive spaces, welding, and deep sea or war zones. Robots can operate in hostile environments like space and planetary locations. The robots can be used in precision focus areas like surgeries and other mechanical work.

The disadvantages of using robots are power consumption, maintenance and repair, smooth retrieval of stored data not comparable to the human mind, and the fact they cannot understand reasonable or non-reasonable tasks or any emotions if not programmed accordingly. Additionally, robots can work fast compared to humans, but a small error in the machine may lead to a delay unless fixed by a human [17].

7.5.4 ROBOTS IN THE HOSPITALITY AND TOURISM INDUSTRY

The tourism and hospitality industry has been a forerunner in adopting the last trends to attract guests' attention and refine their focus toward guest satisfaction. One of the prime examples is Henn na Hotel of Japan, recognized for being the first robot-staffed hotel that utilizes the facility in multiple areas, including reception and guest-facing areas. Hilton uses an AI-based intelligent concierge that was industrialized in alliance with IBM.

World's first Watson-enabled concierge "Connie" (Figure 7.10) [18] is proficient at intermingling with people by greeting and having a conversation with them, utilizing the technology of speech recognition and responding effectively to the queries and learning from each communication to enhance its capability to interact

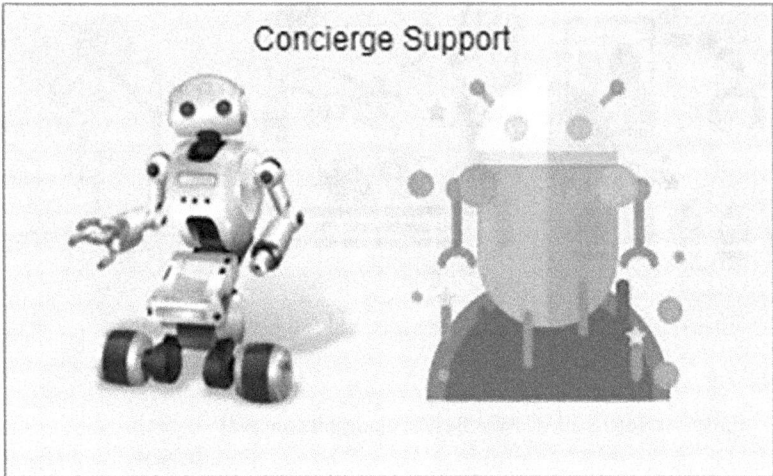

FIGURE 7.10 World's first Watson-enabled concierge "Connie."

Source: IBM Watson Blog [18].

continuously. The intelligent suitcase can self-drive to follow its owner's path and has features to avoid collisions autonomously utilizing collision detection technology and 360-degree turning capabilities.

The deployment of the robotic assistant in the hotel airports and events has fundamentally improved the experience on the guidance related to directions, room status, delivering automatic keys and issuing the entry passes through the capability of robots to use multiple languages and dialects. The intervention of the robots in the market has also attracted guests and visitors who want to understand the operations of such devices an attraction or as a source of entertainment. Robots have been used in multiple areas to entertain clients and engage their idle time. The interaction of a robot in a travel agency may gather key intelligence about customer choices and preferences to support travel agents with a detailed understanding and appropriate client requirements.

Robots have been functional in the security services at the airport to trace concealed objects, weapons and other prohibited items on the flight, along with multiple other services such as luggage porters and butlers, and handling check-ins and check-outs.

7.5.5 ROBOTS IN THE KITCHEN

Moley is the world's first robotic kitchen featuring an advanced, fully functional robot integrated into a beautifully designed, professional kitchen. It has the culinary repertoire of a master chef and can prepare any food item. The consumer version has an iTunes-style library of recipes, all created by master chefs. A pair of articulated robotic hands reproduces the intact purpose with better accuracy, sensitivity and speed. Each move and step can be recorded to prepare a dish, and such accurate operation opens the pathway and confidence for the future market. Imagine that Moley started to prepare a delicious meal that will be ready when you arrive home on your way home. The type of food can be chosen from an IT technology library on your way home. When you are at home, a delicious meal is awaiting you. It starts with a plate of ingredients. The culinary world will see a massive transformation with 3D recipe recording. Think about the employee who arrives home late every night and doesn't have anyone to cook a delicious home meal. Moley will be launched as the new consumer version of the robotic kitchen. The four key aspects will be integrated as kitchen items, for example, robotic arms, oven, and touchscreen unit. The kitchen can be functioned by a touch screen or via a smartphone. In robotic use, glass screens glide across the unit, enclosing it for safe use when no one is home.

Moley captures the imagination of many sectors and could even assist in training future chefs in culinary schools. This robot is multifunctional, and it can hold a knife, chop vegetables, pour sauces, and carry ingredients and food [19]. Moley has 129 sensors which are tactile, contact and proximity sensors. The robot uses contact sensors when holding items [19].

7.5.6 ROBOTS IN FOOD AND BEVERAGE

Robots are in every sphere of the industry. The Nino is an excellent example of the same. An Italian company, Mark and Shakr, launched a cocktail maker widely accepted and used in cruises and hotels. It can make an infinite variety of cocktails and receives orders from guests through smartphones [20].

Another vital step of robots in the industry can be sensed through the innovative launch of Robo Café in Dubai, which offers coffee to the guests in the mall and proves the future has arrived with 100% organic and high-quality coffee [21]. The pioneer in installing robots in the hospitality industry Henn na hotel has a similar setup for the bar as illustrated under. The Henn na bar is a fully robotic run bar where the cocktails and beer are served in different stations and omits the need for humans. The immersive growth of robotics in the industry worldwide proves the importance and rising need for robots. The hotel industry can go entirely on service automation, housekeeping, cleaning, laundry, dishwashing, ordering companion, fully mobile telepresence, interactive booth attendants and robot guides in airports, car rentals, travel agencies, museums and hotels [22].

7.6 INFORMATION AND COMMUNICATION TECHNOLOGY IN TOURISM AND HOSPITALITY

ICT combines continuously increasing communication technologies in the digital world. ICT has impacted the industry for many years with the latest trends like radio, landline phones, television, computers, mobile and smartphones. As stated by UNESCO's glossary, the definition of ICT is a set of technological tools and resources used to transmit, store, create, and share information through telephony, the Internet, and live or recorded broadcasting [23].

7.6.1   ICT and Related Tools

The website, blog, email, live casting, radio or video telecasts, podcasts, satellite and video conferencing are a few examples of the increasing usage of ICT. The concept may be clarified by adjusting the term to information communication and technology [24].

7.6.2   Concept and Components of ICT

The component of ICT is a combination of all the technologies used in the digital world for communication and interaction. Software, hardware transactions, data, cloud computing and communication technology are utilized for societal interpersonal and economic transactions and interactions [25].

The concept of ICT is well explained in Figure 7.11, encompassing all interpersonal social and economic interactions and transactions using several media of communication, including telephone, smartphone, apps, television, or radio to construct the spectrum of the ICT.

In other words, the digital and analog tools used for the information telecast live, recorded video or audio, on the application, or stored in the drive when used in the context of interaction and transaction are dealt with in the ICT branch.

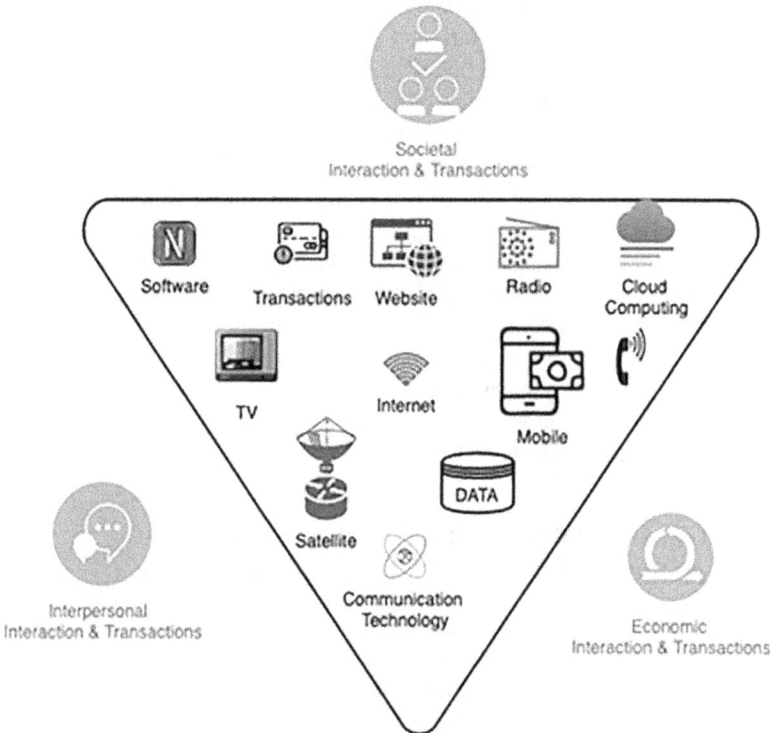

FIGURE 7.11   Information communication technology.

The ICT components are data, software hardware, people, procedures and information. Data is the raw content, and the information is the meaningful output from the shared and cast data.

7.6.3 ICT and Tourism Industry

One of the prime focuses of the UN Sustainable Development Goals is to maximize the reach of ICT in all countries still not in a position to access the Internet by 2020. With its vast influence, the industry has adjusted from non-digital format to entirely or partially digital form in the last few decades due to undue advantages and convenience. A UN agency specializing in information and communication technology named ITU is an official source of global ICT growth support and trends setting [26]. They support underdeveloped countries and economies that need support to access ICT widely.

The tourism industry has transformed itself with the changing technologies [27]. With the availability of the Internet, it has started online travel agencies. The web-booking and check-in facility was provided with easier Internet access to almost everyone in developed countries. Moreover, with the ultra-growth of smartphones and Internet bundles, the applications and information flow have been significant throughout tourism agencies, hotels and airlines. The newly launched businesses were fully strategized for online strategies. At the same time, digital information utilization was also adopted by destination marketing organizations and authorities to complement the usage growth and stress-free convenient online reservation of airlines, the booking of hotels and even taxi booking.

The crucial impact of ICT and IoT has transformed the traditional services of tour guiding, and travel agency has significantly changed their packaging [28]. For example, tour operations have moved to the online booking website. The airline reservation counter has moved to the airline application, and tour guiding does not require a physical tour guide due to virtually available information based on the geo-position of the guest. Airports have started utilizing the ICT in all possible ways to control traffic arrival and departure, and almost all services have been or are fully automatized [29]. Similarly, the hotels have used ICT technologies for room management housekeeping and personalized tablet or television for the rooms [30]. The B2B has also moved fully on the online trend. This proves the indicative purpose of the influence of ICT on the tourism industry, which gave birth to e-tourism. ICT has been the driver of cost-efficient, sophisticated, accurate information and fair competition became possible with the government's support.

Figure 7.12 depicts the areas affected through the ICT and indicates the industry's transformation in all the levels of user business models, delivery, and associated platforms.

7.6.4 Growth Development and Implementation Trend

The trend has risen initially with the ICT growth. Now the industry is entirely dependent on ICT. All the technologies joining the ICT community, like IoT artificial

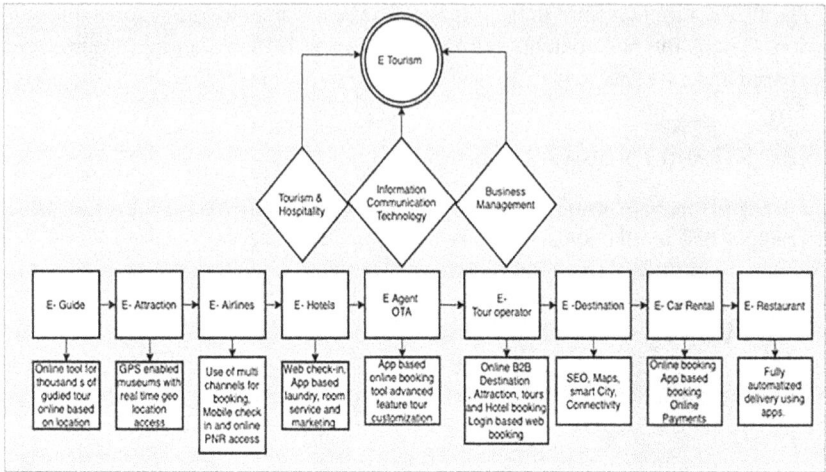

FIGURE 7.12　E-tourism due to ICT.

intelligence involvement of cobots in the hospitality industry, have transformed tourism and hospitality from the vocational field to an innovative technically driven business education vertical. The immediate solution of ICT and the adherence to supportive government policies and industry outlook is proving successful. COVID-19 has impacted lives worldwide; virtual reality has successfully entertained and engaged tourists, guests and visitors.

7.6.5　LIMITATIONS AND CONCLUSION

ICT robotics and other technology are wonderful to implement for convenience, efficiency, and cost-effectiveness. In today's generation, the preferred business model is online using apps. In such an environment, ICT will be a definite tool to reach to palm and mind of each individual with customized packaged products based on his/her choices is the most affluent way to do business. However, transaction security, frauds and unwanted safety issues are always concerns. Another important technology, blockchain, will be explained in detail to utilize for future business models of tourism and hospitality.

REFERENCES

[1]. T. text provides general information S. assumes no liability for the information given being complete or correct D. to varying update cycles and S. C. D. M. up-to-D. D. T. R. in the Text, "Topic: Tourism worldwide," *Statista*. www.statista.com/topics/962/global-tourism/ (accessed Oct. 30, 2020).

[2]. "Top 8 Technology Trends for 2020," *Simplilearn.com*, Aug. 08, 2018. www.simplilearn. com/top-technology-trends-and-jobs-article (accessed Oct. 30, 2020).

[3]. S. P. Mohanty, U. Choppali, and E. Kougianos, "Everything you wanted to know about smart cities: The Internet of things is the backbone," *IEEE Consum. Electron. Mag.*, vol. 5, no. 3, pp. 60–70, 2016, doi:10.1109/MCE.2016.2556879.

[4]. "History of smart cities: Timeline," *Verdict*, Feb. 28, 2020. www.verdict.co.uk/smart-cities-timeline/ (accessed Oct. 30, 2020).

[5]. "Hospitality 2.0 or the role of innovation in the post-covid hospitality industry," *BW Hotelier*. http://bwhotelier.businessworld.in/article/Hospitality-2-0-or-the-role-of-innovation-in-the-post-covid-hospitality-industry/21-09-2020-322900 (accessed Oct. 30, 2020).

[6]. Partner, "What Is Innovation?," *Digital Intent*, Oct. 01, 2018. http://digintent.com/what-is-innovation/ (accessed Oct. 31, 2020).

[7]. G. Satell, "The 4 types of innovation and the problems they solve," *Harvard Business Review*, Jun. 21, 2017.

[8]. "Top 10 IoT applications in 2020 — Which are the hottest areas right now?," https://iot-analytics.com/top-10-iot-applications-in-2020/ (accessed Oct. 31, 2020).

[9]. "Dubai Electricity & Water Authority (DEWA) | Set your AC to 24°C." www.dewa.gov.ae/en/about-us/sustainability/lets-make-this-summer-green/set-your-ac-to-24 (accessed Nov. 03, 2020).

[10]. "AI-powered bin helps restaurants cut food waste | E&T Magazine." https://eandt.theiet.org/content/articles/2019/03/ai-powered-bin-helps-restaurants-cut-food-waste/ (accessed Nov. 03, 2020).

[11]. "IoT & Big data—Sagrada Familia | d-LAB—MWC," *d-LAB*. https://d-lab.tech/project-1/ (accessed Nov. 03, 2020).

[12]. A. Verma and V. Shukla, "Analyzing the influence of IoT in tourism industry," *SSRN Electron. J.*, 2019, doi:10.2139/ssrn.3358168.

[13]. "What are the main types of robots?," *RobotWorx*. /faq/what-are-the-main-types-of-robots (accessed Nov. 01, 2020).

[14]. "Catalyst_Manufacturers_Guide_To_Robotics_-updated-Jan2019.pdf." www.catalyst-connection.org/wp-content/uploads/2019/01/Catalyst_Manufacturers_Guide_To_Robotics_-updated-Jan2019.pdf (accessed Nov. 02, 2020).

[15]. "Blog-Cobot-Basics-4-Types-of-Collaborative-Robots-for-Industrial-Applications-1024x499.jpg (1024×499)." www.elixirphil.com/wp-content/uploads/2020/02/Blog-Cobot-Basics-4-Types-of-Collaborative-Robots-for-Industrial-Applications-1024x499.jpg (accessed Nov. 02, 2020).

[16]. "[Part 4] Cobot basics: 4 types of collaborative robots for industrial applications | Blog | Industrial equipment supplier—Elixir Philippines." www.elixirphil.com/part-4-types-collaborative-robots-industrial-applications/ (accessed Nov. 02, 2020).

[17]. "Advantages and disadvantages of using robots in our life | Science online." www.online-sciences.com/robotics/advantages-and-disadvantages-of-using-robots-in-our-life/ (accessed Nov. 02, 2020).

[18]. "The world's first Watson-enabled hotel concierge robot—IBM Watson," *Watson Blog*, Mar. 09, 2016. www.ibm.com/blogs/watson/2016/03/watson-connie/ (accessed Nov. 02, 2020).

[19]. "Moley builds robots that cook for you," *TechDrive*, Apr. 17, 2015. https://techdrive.co/moley-builds-robots-that-cook-for-you/ (accessed Nov. 02, 2020).

[20]. "Nino robotic bartender can make 'any drink in seconds,'" *Dezeen*, Jun. 07, 2018. www.dezeen.com/2018/06/07/nino-robotic-bartender-can-make-any-drink-in-seconds/ (accessed Nov. 02, 2020).

[21]. "RoboCafe – 100% organic high quality coffee." www.robocafe.ae/ (accessed Nov. 02, 2020).

[22]. S. Ivanov, C. Webster, and K. Berezina, "Adoption of robots and service automation by tourism and hospitality companies," p. 18.

[23]. "Information and communication technologies (ICT)," Jun. 22, 2020. http://uis.unesco.org/en/glossary-term/information-and-communication-technologies-ict (accessed Nov. 02, 2020).

[24]. J. Giles, "What is ICT? What is the meaning or definition of ICT?," *Michalsons*, Nov. 05, 2018. www.michalsons.com/blog/what-is-ict/2525 (accessed Nov. 02, 2020).

[25]. "What is ICT (Information and Communications Technology)?," *SearchCIO*. https://searchcio.techtarget.com/definition/ICT-information-and-communications-technology-or-technologies (accessed Nov. 02, 2020).

[26]. R. ITU, "International telecommunication union," *www.itu.int*, 2020. www.itu.int/en/ITU-D/Statistics/Pages/about.aspx (accessed Nov. 02, 2020).

[27]. A. Verma, V. K. Shukla, and R. Sharma, "Convergence of IOT in tourism industry: A pragmatic analysis," in Journal of Physics: Conference Series, Vol. 1714, No. 1, 2021, p. 012037, IOP Publishing, Goa, India.

[28]. V. Amit, and S. Vinod, Analyzing the Influence of IoT in Tourism Industry, Feb. 23, 2019 (Proceedings of International Conference on Sustainable Computing in Science, Technology and Management (SUSCOM), Amity University Rajasthan, Jaipur—India, February 26–28, 2019). https://ssrn.com/abstract=3358168 or https://doi.org/10.2139/ssrn.3358168

[29]. A.L. Madana, V. Kumar Shukla, R. Sharma and I. Nanda, "IoT enabled smart boarding pass for passenger tracking through Bluetooth low energy," in 2021 International Conference on Advance Computing and Innovative Technologies in Engineering (ICACITE), 2021, pp. 101–106. IEEE, Noida, India, doi:10.1109/ICACITE51222.2021.9404602.

[30]. D.D. Albuquerque, V.K. Shukla, A. Verma, S.K. Tyagi and P. Sharma, "Enhancing sustainable customer dining experience through QR code and Geo-fencing," in 2020 International Conference on Computation, Automation and Knowledge Management (ICCAKM), 2020, pp. 190–196. IEEE, Dubai, UAE, doi:10.1109/ICCAKM46823.2020.9051470.

Part III

Computational Intelligence
for Intelligent Transport
and Communication

8 A Study of Intelligent Transportation Systems

Raja Kunal Pandit and Rahul Nijhawan

CONTENTS

8.1 INTRODUCTION

Surface transportation systems confront a double problem regarding adverse weather on the roads. Due to the reduction in total surface transportation productivity, the financial results of road weather patterns are remarkable. Road weather events significantly increase travel time and accident probability, impairing surface transportation system reliability [1]. Moreover, travelers have to cancel or reschedule their

DOI: 10.1201/9781003218715-11

discretionary trips to avoid the adverse consequences of lousy road weather, which negatively impacts regional economic activities. The traffic demands are usually far lower than they used to be on average, but the crash frequency increases significantly [2–4]. According to research on the intensity of severe road weather conditions associated with crashes during snowfall, cautious driving lowers the seriousness of accidents but not the rate.

Surprising climate patterns also contribute to dangerous road conditions; for example, heavy rain causes poor visibility and flooding [5–6]. The impact of road weather patterns on the occurrence and intensity of crashes has long been regarded as a critical goal of traffic management; understanding the effects of road weather events on both the severity and frequency of crashes is critical [7]. Most road climate accidents occur in the South and Midwest, with rain being the primary cause in the South and snow being the chief reason in the Midwest [8]. Road-climate–related collisions impose a high economic cost on society. Snowy circumstances account for about 25% of total traffic accidents in Finland, with similar conditions in other countries [9–10].

Remarkable weather drift presents a bleak view of global weather in the future, with historical weather models predicting more frequent weather extremes [11–12]. As a result, all-weather conditions affecting surface transportation operations are likely to become more extreme and regular. As predicted, periodic events disrupt global passenger and freight transit, and also impact the transportation system [13–14]. Increased storm surges will put coastline surface transportation networks at risk, resulting in repeated floods [15–16]. Transportation agencies have been able to design innovative solutions related to the transportation difficulties of the 21st century [17]. The intelligent transportation system has been a critical component in mitigating the effects of bad road weather on traffic operations from the beginning of technology in transportation [2] [18].

Transport administrators can use various traffic operations programs, especially winter maintenance programs, to provide travel-related information, which will be useful in typical traffic operations [3]. In order to minimize the problem during the road trip, ITS can be used to create dynamic road weather-informed trip regulations that deliver correct route information to travelers and shippers in a timely manner. For the sake of highway safety and trip reliability, a strong surface transportation infrastructure management program will need to enhance ITS-based creative solutions in the future to handle the challenges of more unfavorable road weather patterns.

8.2   ITS INFRASTRUCTURE FOR ROAD CLIMATE MANAGEMENT: PRINCIPLES AND APPLICATIONS

According to a Transport Research Board assessment, adverse weather will impact 57% of the US population who reside close to the coast [19]. The following are the principal modes of road transportation that are affected by weather:

1. Coastline highways have been repeatedly flooded due to rising sea levels.
2. The load-carrying ability of bridges and pavement is reduced when the number of hot days increases over a year.

3. Precipitation intensity, in the form of rain or snow, obstructs traffic operations and jeopardizes traffic safety.
4. The load on emergency escape management grows as the recurrence of tornadoes and hurricanes rises.

ITS technology gives timely information and guidance [19] to minimize the impact of projected future climate on terrestrial transportation systems. The bad judgment of the driver [20–21] is the primary cause of road accidents in adverse climates. Transportation organizations provide travelers with helpful roadway statistics such as advised limits and road surface data to prevent driver-related mistakes. Reducing driving conditions under harsh road conditions significantly reduces safe driving speeds, and most drivers are unaware of the dangers, increasing the crash risk [22]. Transportation agencies depend on the ITS infrastructure to acquire and anticipate road weather conditions and notify travelers about safe driving speeds and other important traveler information via variable message signs (VMSs) and web portals [23]. For real-time trip planning, accurate and timely data is critical [24].

The most common use of ITS infrastructure is to alert drivers about present and prospective traffic conditions, which aids in route planning and optimization [25]. Dynamic message signs such as smartphone apps, advisory highway radio, and social networks are used to provide information to travelers. Kilpelainen and Summala conducted an on-site survey and discovered that travelers who acquire road weather information are more willing to postpone their visits to avoid seriously impacted routes. On the other hand, travelers acknowledge that real driving situations were better than projected and that nighttime road conditions were worse [23]. Actual road conditions, not road weather predictions, influenced route driving habits the most. Localized road weather predictions increased driver behavior to improve safety in severe road weather using ITS applications [26–28]. ITS infrastructure applications for road weather management:

8.2.1 ROADWAY CONSERVATION

Winter road restoration accounts for many transportation agencies' annual maintenance budgets [1]. The Michigan Department of Transportation (DOT) spends 40% of its overall expenditures during winter [29]. In many countries, a similar situation exists [10]. Because the efficiency of planned maintenance relies on minimal resources to recover traffic volume and ensure driver safety conditions, the quick and accurate availability of real-time road weather information is critical. However, traffic's vast network of roads tends to not interpret road weather according to local weather. As a result, the inaccuracy of such forecast information reduces the effectiveness of the maintenance program's proactive planning and causes a significant increase in the DOT maintenance budget. Installing a road weather information system (RWIS) at strategic locations will provide more localized and accurate information. This allows transport agencies to perform proper maintenance operations [30–31].

Using winter repair operations, federal agencies are combining automated weather systems for collecting localized road weather and traffic updates [29]. The local climate, particularly the temperature at the highway level, does not correctly anticipate subtle changes. International transportation bodies acknowledge this high-way weather problem. The Federal Highway Administration (FHWA) in the United States has completed a set of pilot initiatives to help the Seasonal Maintenance Program, backed by ITS infrastructure and the Maintenance Decision Support System (MDSS), build several weather sources cooperatively [32]. The winter road climate maintenance strategy is the most effective and localized. A related DSS was created to aid in implementing and scheduling climate-responsive traffic manage-ment operations.

DOTs in the United States have claimed significant benefits from MDSS imple-mentation, including lower salt and other reagent consumption, improved equip-ment utilization, and reduced labor [33–34]. Furthermore, environmental models and road temperatures, a road weather platform, are used by Russia, Slovakia, and Sweden to estimate real-time highway surface conditions for wintertime mainte-nance strategies [35].

The United Kingdom has erected several road climate radar systems to collect real-time radar images and build maintenance procedures, and the planned mainte-nance has improved [36]. For wintertime routine maintenance, Sweden is testing the use of floating vehicle sensor data paired with RWIS data [37]. In the United States, a related firm is being investigated [38]. France has been employing an autonomous data merging architecture, "Optima," which integrates numerous road weather data sources to change road weather patterns every 5 minutes and has shown to be bene-ficial for real-time pavement climate monitoring [39].

8.2.2 REAL-TIME TRAFFIC CONTROL

Efficient traffic control systems provide the capabilities needed to adopt flexible operational techniques depending on various road circumstances, such as traffic and climate conditions. [40]. Traffic simulators are utilized to select the optimum flexible service solutions to analyze existing and expected short-term road circumstances.

Adverse road climate is a major reason for non-traffic congestion in the United States, accounting for 25% of overall delays and 1 billion hours of lost productiv-ity [41]. It is possible to improve vehicular traffic by including the effects of road weather factors in traffic management strategies [1]. Real-time road traffic and weather information gathering via ITS infrastructures are critical for quantifying climate effects [40].

8.3 RESOURCES FOR CLIMATE DATA ON THE HIGHWAY

The effectiveness of climate data and the capacity to compute properly are essential for effective traffic management amid road climate threats [42]. However, providing consumers with consistent road climate data continues to be a difficulty, which is presently hampered by the following:

- Transport system networks and climatic variation spatial variability.
- There are no surveillance systems in place to gather road climate data. Infor-mation on road climate is critical to a successful solution.

8.3.1 Resources of Accessible Data

Logistics companies rely heavily on data provided by the government. On the other hand, official weather forecasts aren't specifically specialized for highway climate monitoring. As a result, RWIS and private road weather prediction services are used by government authorities. RWIS deployment for vast highway networks is hampered by higher development and maintenance expenses.

RWIS is maintained by most state DOTs in the United States along strategic regional highway crossings to formulate roadway repair strategies and alert motorists ahead of time. Across the globe, most RWIS terminals have been utilized to compile meteorological characteristics such as ambient temperature, wind speed, roadway condition, air quality, radiation intensity and surface friction coefficient [43]. DOTs build an acknowledgment plan for highway maintenance based on the data collected and inform travelers via DOT websites, traffic hotlines (511), radio channels and VMSs [43].

European countries collaborate to create guidelines to assure RWIS deployment consistency and compatibility [44]. Using ITS technologies to combine numerous climate sources on a single platform could enhance the efficacy of road weather information for judgment (Table 8.1).

TABLE 8.1
ITS Techniques for a Variety of Road Weather Situations

Weather Events	Impacts on Transportation	ITS Applications	Limitations
Rain	Reduces friction and visibility and also reduces travel speed. Increases the risks of flooding and crashes.	Automatic warning system via embedded sensors. Variable message signs.	Embedded intrusive sensors detect the wetness of roads. Non-intrusive sensors detect the presence of rain.
Flooding and Extreme Waterfall	Highways submerge due to frequent flooding. Reduces travel speed.	Intelligent mounted device, automated phone/warning system.	Interoperability between different system components.
Snow	Reduces pavement friction, travel speed and roadway capacity. Increases crash risk and vehicle control difficulty.	Snow detection sensor. Automatic warning system and anti-icing message. Chemical concentration detection sensor.	Selection of treatment for different types of precipitation, i.e., sleet, freezing rain, blowing snow and frost events. To initiate anti-icing treatment.
Tornadoes or High Winds	Highway hindrance reduces access to disaster areas. Reduces visibility.	Automatic message/ warning system, wind sensors.	Limited range of wind sensors, false alarm rates.
Fog/Mist	Increases detours, crash risk, decreases visibility.	Forward scatter sensors, radar speed sensors, transmissometer.	Limited range of sensors requires a larger number of devices to cover short distances.
Temperatures	Reduces load carrying capacity of bridges.	Intelligent temperature probe system, roadside weather stations.	Temperature sensor system deployment cost.

8.3.2 COORDINATION BETWEEN THE PUBLIC AND PRIVATE SECTORS

Effective climate observing systems have been built with private support. Several common datasets, on the other side, predict a vast region unsuitable for developing route-specific information. Because of the volatility and accuracy associated with individual data sources, it is required to combine many sources to obtain more reliable data. For example, we may create a representative group of future possible highway climate occurrences to use an organization forecast. Several meteorological data sources are being used and compiled for surface transport in the country. Such activities will aid in analyzing and developing road weather predictions, allowing them to be used at various levels of transport judgment [38].

Private donors rely on public equipment and climate data to create accurate local predictions for transportation maintenance and operation [45]. Traffic experts now have better forecasting skills because of recent technological advances in collecting road weather information via vehicle-based sensing devices [33]. Sensor-equipped servicing trucks can provide highway weather data near the pavement and can be combined with other climate sources of data [46]. To collect trustworthy roadway weather information, however, wearable sensor technologies must be developed [47].

8.4 ON-ROAD APPLICATION OF ITS TECHNOLOGIES IN DIFFERENT SEASONAL CHANGES

Despite the widespread use of road weather applications in developed countries, the number of applications was very limited in developing countries. In order to use ITS solutions to their maximum potential, developing countries must have well-organized and planned long-term national strategies [48–49].

Furthermore, let us look at several ITS-based solutions to reduce the harsh effects of road weather occurrences in other nations. The table above depicts the consequences of various road weather patterns and existing ITS implementations and their limits.

8.4.1 RAIN

Due to the sheer adverse weather, road conditions are deteriorating. It also reduces vehicle speed [50–52]. Increased collision hazards and reduced roadway capability are caused by decreased roughness and impaired visibility during rain [8]. According to the FHWA, rain is responsible for roughly 47% of climate accidents in the country [53], and similar trends have been documented on other nations' roadways [54].

Autonomous safety systems combined with highway wet sensors, which reduce the average speed of vehicles by 16 km/h and accident frequency in intense rain conditions, have been effectively implemented in rainy regions such as Florida. The Colorado Department of Transportation [55] has erected an innovative climate alert system on state highway 82 in Snowmass Canyon. It assesses highway surface tension utilizing non-intrusive sensors and notifies vehicles via a VMS posted 1 mile ahead with suggested velocity and a caution text.

The expressway stretch did not see any winter accidents since the technology was implemented [56]. Climate change will result in a more extreme downpour in the future [19], necessitating more preemptive measures to ensure the transport system's durability.

8.4.2 FLOODING AND EXTREME WATERFALL

The cities on coastal belts have a significant challenge of rising sea levels. It has been concluded in recent studies that the cities around the world near the sea have a grim future [57–58]. Frequent and extensive flooding always hits network transportation at its most challenging. The addition of global warming patterns in infrastructure development, planning and maintenance are the pillars on which the success of fast recovery depends. For example, in most of New York's coastal areas, most analyses have projected that by 2050, the city will be largely inundated owing to modest sea level rise [59].

Although sea level rise may not pose a considerable difficulty in daily traffic management, the principal consequences will be felt later [59]. According to research [59], the cost of a direct impact by a major storm on New York City may be as high as $100 billion, considering the damage to transportation infrastructure. ITS structures have been used to anticipate high water occurrences. For example, in Palo Alto, California, detectors with lenses monitor water levels. If the sensor detects any flood risk, an automated warning/phone call will be sent to all residents. Simultaneously, emergency and transportation managers may use sensor data to improve emergency plans, boosting the efficiency of public and staff safety [60].

In San Antonio, Texas, an identical system was established to warn passengers of high-water dangers and alert emergency responders [61]. TXDOT implemented an intelligent drainage system pumping system that contributes to a flood risk system to maintain track of genuine pump functions and provide the latest information to highway managers throughout flooding occurrences [61]. In the event of a pump failure, wayside innovativeness signs were triggered, warning of possible flooding. Service providers must closely follow ITS rules to make various systems compatible.

NTC transportation communication service providers have utilized ITS protocol standards to verify that various communication systems are interoperable [61]. Advanced transportation weather information systems (ATWIS) have employed this ITS application. ATWIS has been acknowledged to be beneficial in pre-trip preparation and transit trip planning by travelers [62–63].

8.4.3 SNOW

Over two-thirds of the US roadway system gets at least five inches of snow annually [53]. Snow events account for one-fourth of all highway accidents. During snowfall, UK roadway collisions account for 2.8% of overall crashes, with northern portions accounting for up to 5.9% of total crashes [64]. According to Perry and Symons, road conditions can become hazardous if over 20 mm of ice collects [64]. To prevent snow buildup, a conventional snow treatment includes plowing the snow and spraying anti-icing agents.

Roadway maintenance involves assessing the road conditions in snowy conditions as accurately as possible. Norman et al. used RWIS data gathered on Swedish highways to construct a link between slipperiness and collision risk to determine the most extreme slipperiness [65]. Excessive precipitation was the greatest danger for driving conditions regarding ice on the roadways [66], which greatly reduced surface friction [67].

Maintenance crews can estimate snow accumulation and intensity with sensor-equipped servicing trucks and RWIS, allowing them to move resources more efficiently [30]. Washington DOT conducted a pilot study to determine the efficiency of automated anti-icing chemical agent spraying. A cost-benefit study [30] confirms the importance of automatic anti-icing technology in decreasing snow dangers. In numerous states in the United States and Europe, traffic controls such as adjustable speed restrictions have shown to be an effective strategy for controlling traffic amid severe highway conditions [66] [68–69].

Various routes in Arizona are subjected to various weather risks; for example, a section of Interstate 10 is subjected to dust threats in the springtime and snow/ice concerns in the winter. Excited travelers can sign up for automated email alerts about severe weather along the route [70]. Several intelligent transportation systems (ITS) have improved traffic control during snowstorms.

8.4.4 TORNADOES/HIGH WINDS

Strong winds and tornadoes wreak havoc on assets and livelihoods, providing a significant challenge for transportation infrastructure and road traffic [71]. ITS technology (such as RWIS) was installed along road infrastructure to detect high wind sensitivity circumstances and alert transportation controllers and travelers to plan for adverse road seasonal changes. Authentic roadway data can be gained using advanced mobile probe data (i.e., emergency vehicles) to speed up restoration work and preparations during strong wind season [72–73].

In northeast Florida, for instance, 20 light wind sensors were placed on vital structures. These devices send data to a central control unit via the NOAA Geostationary Operational Environmental Satellite, which reduces recurring expenses and improves real-time communication [74]. Wind sensors are being used to issue dust storm warnings. For example, in southwest New Mexico, a dirt weather alert system was established that includes a wind sensor, a visibility sensor, and CCTV cameras to advise passengers by VMS, advised radio, and the Internet [75].

Countries like the United Kingdom and Scotland have used similar wind-sensor-based traffic warning systems to restrict vehicle access to motorways during high wind speeds [53]. Due to the severity of the wind, certain portions of roadways, particularly bridge sections, may need to be shut down totally [54].

8.4.5 Fog/Mist

Mist is defined as the sudden appearance of thick clouds over the road surface, reducing driver vision to zero [76]. Annually, fog contributes to several multi-vehicle collisions, highway systems, and other highway functional classifications in the United States [27]. In the United States, fog-related highway weather collisions result in an average of 600 fatalities. [77] Transportation authorities built a visibility assessment sensor-based fog alert system at most cloud-prone highway sections to alert passengers about misty road climate and take conditional steps to reduce collision risks [78].

These solutions use sensor visibility requirements to let visitors take care and choose remedial measures like slowing down or choosing alternate routes [79–80]. Following a multi-vehicle incident on the Interstate 10 Bay Bridge in 1995 due to limited visibility, the Alabama DOT integrated reverse sensor technology with CCTV cameras to construct an automated alert system, which reduced fog-related accidents [79]. Necessary safety measures, such as warning signs based on a visible barrier, were remotely managed and communicated with DOT sections and law enforcement agencies [56]. Fog alert systems have been designed in 18 states throughout the United States [80].

However, selecting suitable sensor technology to reduce false alarm rates due to insufficient site-specific fog generation parameters is a significant difficulty. The United Kingdom, Finland, and Saudi Arabia have all implemented similar procedures [76] [80].

8.4.6 Temperature

The rising average temperature is the primary worry of climate change [81]. In warm nations like Saudi Arabia, hot summer temperatures have been identified as a serious concern for the transportation infrastructure [82]. Existing equipment was not built to withstand architectural deformation in extreme heat [83]. High temperatures diminish the load-bearing capacity of roadways, promoting pavement degradation such as rutting and other flaws [84].

An increase in temperature causes highway overpass segments to expand unexpectedly, perhaps leading to expansion failure. To reduce pavement damage during the summer, northern states in the United States have imposed axle and gross vehicle restrictions on bulk trucks. If periodic roadway conveying freight transportation limits are alleged, such restraints will stifle economic growth [85–86]. Devices have been utilized to monitor structure health at important frame structures to offer enhanced load limitation information for bridges under temperature extremes. These sensors will send out notifications to traffic management, prompting them to take corrective action, such as redirecting traffic to alternate pathways or enforcing limits [87].

Additionally, by placing temperature sensors on multiple road pavements, a comprehensive temperature may be obtained, which can be fed into models for estimating highway deterioration. The Alaska DOT, for example, has a huge issue during the spring thaw, which affects pavement load-carrying ability [88–89]. It placed sophisticated temperature data sensors to detect temperatures at various road pavements to determine the maximum axle load and apply seasonal load limitations [90].

All parties, including industrial transport firms, were given 3 days' notice using historical temperature records and national weather projections connected to sensor data [90].

8.5 CHALLENGES AND OPPORTUNITIES FOR FUTURE STUDIES

The core of subsurface transport system highway climate monitoring is ITS architecture. The US Department of Transportation's ITS strategic study design for 2010–2014 called for a comprehensive data gathering structure to improve road climate management practices and support the next phase of dependable highway meteorological solutions [68].

It is critical to make a conscious effort to obtain climate and infrastructure partners. The purpose is to develop an operational strategy that will remove roadblocks to a smooth transition from present restricted roads and bridges highway weather management approach to a vehicles and platform robust highway weather management platform. Additional efforts are required to develop road weather-aware ITS systems, as stated in the following subcategories.

8.5.1 CONNECTIVITY

The compatibility of old and new technologies will be a major concern for transport workers as technology advances. When updating ITS road climate management systems, DOTs face compatibility concerns [18]. Various organizations' climate remote operations generate different types of climate data that can be easily processed to be relevant for highway transport systems [91].

The file type must be consistent and standardized. Several standard-setting organizations in the United States and Europe work together to produce standards to improve the compatibility of ITS network elements. Experts from the public and commercial sectors collaborate in a step-by-step procedure to develop a market-ready required standard [92]. The use of rules in the acquisition of ITS systems can greatly reduce compatibility concerns [93].

8.5.2 COLLABORATION BETWEEN AGENCIES

Most climate detection systems do not gather or forecast severe weather at the earth's surface. For highway climate monitoring, however, ground-level meteorological data is required [18]. The combination of diverse climate sources to create trustworthy road weather information for wintertime maintaining roads has been investigated by federal efforts in the United States, such as the Clarus [38] [94] initiative, on a global scale. After finishing the COST program, the International Weather Commission was established, which has resulted in regional developments in road climate regulation [95]. These collaboration endeavors must be permitted to advance road climate organization studies innovations.

8.5.3 WEATHER TREND INCORPORATION DIFFICULTIES
IN ITS MANAGEMENT AND FORECASTING

The mobility community is increasingly paying much attention to weather incidents, and one such danger is a strong rainfall event, which is difficult to factor into ITS distribution plans. Precipitation and its excesses fluctuate due to climate change [96]. More "strong downpour" occurrences, including thunderstorms, subtropical rain/cyclones, hurricanes, and snowfalls, result from this occurrence. Ultimately, such occurrences point to a greater danger of floods.

8.6 CHALLENGES AND SOLUTIONS

8.6.1 IDENTIFYING AND FORECASTING EXTREME RAINFALL EVENTS

There is growing evidence that climate change causes rainfall events with serious economic consequences [97]. Generally, extreme precipitation outperforms the long-term mean for short moments. Extremes are classified differently depending on the demands of the consumers; therefore, developing easily accessible terminology for highway climate management is crucial for providing better information to the users [12]. Weather extremes must be characterized to facilitate communication within the ITS industry. With the move from rain-gauge to infrared and remotely sensed measurements, a great success in quantifying moisture has been realized.

However, detecting exceptional precipitation at limited geographical scales remains challenging, posing a problem in characterizing and interpreting the data to the demands of consumers. To enhance ITS-assisted road climate control, precise high rainfall values at finer range scales are required. Despite the improved precision of estimated measurements, gauges cannot deliver information at the more appropriate geographical spatial scales [98].

In contrast, decreasing automated process uncertainties during the autonomous method's translation of radiometric observations into rainfall amounts at the surface remains difficult. Due to the spatial complexity of events, such as tropical storms, which can impair traffic operations at a regional level, whereas heavy precipitation impacts traffic at a small level, climate research groups have challenges in modeling weather anomalies [99–103]. Due to flaws in modeling parameters, predicting significant precipitation at a temporal scale is also difficult.

8.6.2 USING EXISTING SEVERE WEATHER DATA TO MINIMIZE
THE RISK OF INFRASTRUCTURE FAILURE

Flooding will become more common due to global warming [104–105]. Seaside flooding is also a threat due to rising sea levels, severe storms, and hurricanes. Because these incidents will cause major disruption to mobility network infrastructure, using this data will improve ITS implementation. Urban floods will significantly impact urban mobility, which is a worry because sewer systems cannot release during high precipitation situations. By 2030, it is expected that over 60% of the world's population will live in urban regions [106], making urban ITS infrastructure crucial in handling the increasing traffic demands in city environments.

The expansion of urban risk of flooding due to increased hydro weather and climate variability and micro-climatic changes generated by urbanization would inevitably be a key issue for ITS [107]. Research has revealed the link between urban sprawl and urban climate in the last two decades. The most rapid urban floods [108] were caused by the "urban heat island" phenomena. As a result, evaluating these occurrences is helpful in preparing urban ITS architecture, and the difficulty continues to:

- Recognize the impact's origin.
- Measure future situations caused by the interaction of several variables, all of which have a larger level of uncertainty.

8.6.3   General Circulation Model (GCM) Role in ITS

Severe storm occurrences are becoming more frequent as a result of climate change. As a result, expected weather and climate variations will greatly impact road weather-aware ITS. In a preliminary phase, it's critical to assess global warming and climate change and how they can affect ITS network design. The usage of IPCC-proposed GCMs [109–110] has been broadened to give anticipated scenarios.

The statistical geostatistical technique has frequently been used to scale down climatic data to a local scale to utilize GCM information. Designed to simulate precipitation for future climatic scenarios is difficult since early discussions, parameterized patterns, and interdependencies are all important. However, as GCMs have evolved, performance in terms of forecast rainfall volume, severe events, and atmospheric circulation rainfall variation has enhanced [109–112]. This improved climatic data can potentially be used in ITS architecture development and operation.

8.7   CONCLUSION

Whereas transportation companies increasingly depend on ITS infrastructures for significant traffic surveillance and management, the full capability of ITS for controlling surface transport networks is still to be realized. More research is needed to clarify the complex interplay between detrimental road weather patterns and traffic movements. This will help to identify the effective characteristic features of serious road weather patterns and preemptively estimate the risk associated with them by trying to incorporate transportation reliability of the system, adaptability and frailty.

The accuracy of weather forecasting techniques has improved over time. New satellite surveillance and weather forecasting techniques have significantly improved and will continue to do so. However, path road climate data, such as the likelihood of individual routes flooding, is still difficult to come by. Obtaining highway weather data necessitates enhancing the quality of the climate data gathering system to assess the selected weaknesses of broad highway networks, something existing infrastructure cannot achieve. It will be feasible to supplement the existing lack of path road weather data using portable road climate data obtained by linked automobiles.

Cellular services climate sensors, a vast mobile network analysis ability and efficient management information systems must all be developed to effectively use future vehicular communication road climate information management architecture. Emerging techniques must be standardized to match historical ITS and highway weather information-gathering systems to increase the accuracy of road weather information by combining data from numerous sources and reducing the risks that come with information from a data origin.

REFERENCES

[1]. T. H. Maze, M. Agarwal, and G. Burchett, "Whether weather matters to traffic demand, traffic safety, and traffic operations and flow," Transp. Res. Rec., J. Transp. Res. Board, no. 1948, pp. 170–176, 2006.

[2]. M. Agarwal, T. H. Maze, and R. Souleyrette, "Impacts of weather on urban freeway traffic flow characteristics and facility capacity," in Proc. of the Mid-Continent Transp. Res. Symp., Ames, Iowa, USA, August 2005, pp. 1–20. © 2005 by Iowa State University.

[3]. M. Cools, E. Moons, and G. Wets, "Assessing the impact of weather on traffic intensity," in Proc. Annu. Meet. Transp. Res. Board, Washington, DC, USA, 2008, [CD-ROM].

[4]. A. De Palma and D. Rochat, "Understanding individual travel decisions: Results from a commuters survey in Geneva," Transportation, vol. 26, no. 3, pp. 263–281, 1999.

[5]. M. Agarwal, T. H. Maze, and R. Souleyrette, "The weather and its impact on urban freeway traffic operations," in Proc. 85th Annu. Meet. Transp. Res. Board, Washington, DC, USA, Transportation Research Board of the National Academies, 06–1439, 2006, [CD-ROM].

[6]. S. Datla, and S. Sharma, "Impact of cold and snow on temporal and spatial variations of highway traffic volumes," J. Transp. Geography, vol. 16, no. 5, pp. 358–372, 2008.

[7]. D. Eisenberg, and K. Warner, "Effects of snowfalls on motor vehicle collisions, injuries, and fatalities," Am. J. Public Health, vol. 95, no. 1, pp. 120–124, 2005.

[8]. P. A. Pisano, L. C. Goodwin, and M. A. Rossetti, "U.S. highway crashes in adverse road weather conditions," in Proc. 24th Conf. IIPS, New Orleans, LA, USA, 2008.

[9]. R. Kulmala, and P. Rämä, "The effects of weather and road condition warnings on driver behavior," in Proc. Conf. Road Safety Eur. SHRP, Prague, Czech Republic, pp. 169–178, 1995.

[10]. A. K. Andersson, Winter road conditions and traffic accidents in Sweden and U.K.: Present and future climate scenarios, M.S. thesis, Univ. Gothenburg, Gothenburg, Sweden, 2010.

[11]. C. Mora et al., "The projected timing of climate departure from recent variability," Nature, vol. 502, pp. 183–187, 2013.

[12]. IPCC, Managing the Risks of Extreme Events and Disasters to Advance Climate Change Adaptation, A Special Report of Working Groups I and II of the Intergovernmental Panel on Climate Change [C.B., Field, et al. (eds.)]. Cambridge University Press, Cambridge, and New York, 2012.

[13]. F. Curtis, "Peak globalization: Climate change, oil depletion and global trade," Ecological Econ., no. 69, pp. 427–434, 2009.

[14]. M. J. Koetse, and P. Rietveld, "The impact of climate change and weather on transport: An overview of empirical findings," Transp. Res. Part D, vol. 14, pp. 205–221, 2009.

[15]. V. Gornitz, "Sea-level rise and coasts," in C. Rosenzweig and W. D. Solecki (eds), Climate Change and a Global City: The Potential Consequences of Climate Variability and ChangeMetro East Coast. Columbia Earth Institute, New York, 2001.

[16]. L. M. Dingerson, Predicting future shoreline condition based on land use trends, logistic regression, and fuzzy logic, M.S. thesis, Virginia Inst. Marine Sci., College of William and Mary, Gloucester Point, 2005.

[17]. R. P. Maccubbin et al., "Intelligent transportation systems benefits, costs, deployment, and lessons learned: 2008 update," in Federal Highway Administration, Washington, DC, USA, 2008.

[18]. L. Zhang, J. Colyar, P. Pisano, and P. Holm, "Identifying and assessing key weather-related parameters and their impacts on traffic operations using simulation," in Proc. 84th Annu. Meet. Transp. Res. Board, Washington, DC, USA, Transportation Research Board of the National Academies, 05–0962, 2005, [CD-ROM].

[19]. Transportation Research Board (TRB), The Potential Impacts of Climate Change on U.S. Transportation, National Academy of Sciences, Washington, DC, USA (Special Rep. 290), 2008.

[20]. E. Petridou, and M. Moustaki, "Human factors in the causation of road traffic crashes," Eur. J. Epidemiology, vol. 16, no. 9, pp. 819–826, 2000.

[21]. L. Evans, "The dominant role of driver behavior in traffic safety," Am. J. Public Health, vol. 86, no. 6, pp. 784–786, 1996.

[22]. J. B. Edwards, "Speed adjustment of motorway commuter traffic to inclement weather," Transp. Res. Part F, Traffic Psychol. Behaviour, vol. 2, no. 1, pp. 1–14, 1999.

[23]. M. Kilpeläinen, and H. Summala, "Effects of weather and weather forecasts on driver behavior," Transp. Res. Part F, Traffic Psychol. Behaviour, vol. 10, no. 4, pp. 288–299, 2007.

[24]. E. Urbina, and B. Wolshon, "National review of hurricane evacuation plans and policies: A comparison and contrast of state practices," Transp. Res. Part A, Policy Practice, vol. 37, no. 3, pp. 257–275, 2003.

[25]. US Dept. Transp., Intelligent Transportation Systems for Traveler Information: Deployment Benefits and Lessons Learned, Washington, DC, USA, 2013 [Online]. http://ntl.bts.gov/lib/jpodocs/brochure/14319_files/14319.pdf

[26]. M. Eriksson, and J. Norrman, "Analysis of station locations in a road weather information system," Meteorol. Appl., vol. 8, no. 4, pp. 437–448, 2001.

[27]. L. C. Goodwin, "Best practices for road weather management Version 2," in Federal Highway Administration, Washington, DC, USA, 2003.

[28]. M. Bayly, B. Fildes, M. Regan, and K. Young, "Review of Crash Effectiveness of Intelligent Transport Systems," Traffic Accident Causation in Europe, Project No. 027763, 2007.

[29]. J. Foley, W. Thompson, and R. Warren, "The use of the 85 percentile speed data as a measure of winter maintenance performance," Presented at the National Rural Intelligent Transportation Society Conf., Seaside, OR, USA, 2009.

[30]. R. Stowe, "A benefit cost analysis of intelligent transportation system applications for winter maintenance," Presented at the Transportation Research Board Annual Meeting, Paper No. 01–0158, 2001.

[31]. U. Farooq, M. A. Siddiqui, L. Gao, and J. L. Hardy, "Intelligent transportation systems: An impact analysis for Michigan," J. Adv. Transp., vol. 46, no. 1, pp. 12–25, 2012.

[32]. National Center for Atmospheric Research (NCAR) The Maintenance Decision Support System (MDSS) Project, Technical Performance Assessment Report, Colorado Field Demonstration, Winter 2007–2008, Boulder, CO, USA, 2008.

[33]. Z. Ye, C. Strong, X. Shi, and S. Conger, "Analysis of maintenance decision support system (MDSS) benefits & costs," in Western Transportation Institute & Iteris, Inc., Bozeman, MT, USA, Final Rep., 2009.

[34]. T. McClellan, P. Boone, and M. A. Coleman, Maintenance Decision Support System (MDSS): Indiana DOT Statewide Deployment Report, Indiana Department of Transportation, Indianapolis, IN, USA [Online]. www.meridian-enviro.com/mdss/pfs/files/MDSSReportWinter08–09.pdf

[35]. S. Karanko, I. Alanko, and M. Manninen, "Integrating METRo into a winter maintenance weather forecast system covering Finland, Sweden and Russia," in Proc. 16th Int. Road Weather Conf., Helsinki, Finland, 2012, pp. 1–5.

[36]. A. H. Perry, L. Symons, and A. Symons, "Winter road sense," Geographical Mag., vol. 58, pp. 628–631, 1986.

[37]. T. Gustavsson, J. Bogren, A. Johansson, P. Ekström, and M. Andersson, "BiFi—Bearing information through vehicle intelligence," in Proc. 17th Int. Road Weather Conf., Andorra, Spain, 2014, pp. 1–6.

[38]. RITA, Clarus, Federal Highway Administration, Washington, DC, USA, 2013 [Online]. www.its.dot.gov/clarus/

[39]. O. Coudert, L. Bouilloud, and A. Foidart, "Optima (Road Weather Informations dedicated to road sections)," in Proc. 16th Int. Road Weather Conf., Helsinki, Finland, 2012, pp. 1–7.

[40]. A. A. Kurzhanskiy and P. Varaiya, "Active traffic management on road networks: A macroscopic approach," Philosoph. Trans. Roy. Soc. A, Math., Phys. Eng. Sci., vol. 368, no. 1928, pp. 4607–4626, 2010.

[41]. R. M. Alfelor and C. Y. D. Yang, Managing Traffic Operations during Adverse Weather Events. Federal Highway Administration (FHWA), Washington, DC, 2011.

[42]. J. E. Thornes and D. B. Stephenson, "How to judge the quality and value of weather forecast products," Meteorol. Appl., vol. 8, pp. 307–314, 2001.

[43]. C. Eng, K. Mahoney, T. Masternak, and J. Vavra, Sharing of Observations Collected with Road Weather Information System Environmental Sensor Stations. Nat. Ocean. Atmosp. Admin. (NOAA), US Dept. Transp. (USDOT), Washington, DC, USA, 2004.

[44]. P. A. Brodard, "EN15518—A European standard for Road Weather Information Systems (RWIS)," in Proc. 16th Int. Road Weather Conf., Helsinki, Finland, 2012.

[45]. V. Nurmi et al., "Economic value of weather forecasts on transportation—Impacts of weather forecast quality developments to the economic effects of severe weather," Seventh Framework Programme, Delft, The Netherlands, EWENT Rep. D5.2, 2012.

[46]. S. Drobot et al., "Improving road weather hazard products with vehicle probe data," Transp. Res. Rec., J. Transp. Res. Board, vol. 2169, no. 1, pp. 128–140, 2010.

[47]. W. P. Mahoney, and J. M. O'Sullivan, "Realizing the potential of vehicle based observations," Bull. Amer. Meteorol. Soc., vol. 94, pp. 1007–1018, 2013.

[48]. K. Chen, and J. C. Miles, "ITS handbook 2004: Recommendations from the world road association," in Permanent Int. Assoc. Road Congr. (PIARC), Paris, France, 2004.

[49]. A. A. Shah, and L. J. Dal, "Intelligent transportation systems in transitional and developing countries," IEEE Aerosp. Electron. Syst. Mag., vol. 22, no. 8, pp. 27–33, 2007.

[50]. F. L. Hall, and D. Barrow, "Effects of weather and the relationship between flow and occupancy on freeways," Transp. Res. Board, Washington, DC, USA, Transp. Res. Rec. 1194, pp. 55–63, 1988.

[51]. W. Brilon and M. Ponzlet, "Variability of speed-flow relationships on German Autobahns," Transp. Res. Board, Washington, DC, USA, Transp. Res. Rec. 1555, pp. 91–98, 1996.

[52]. J. Asamer, and M. Reinthaler, "Estimation of road capacity and free flow speed for urban roads under adverse weather conditions," in Proc. 13th In. IEEE Annu. Conf. Intell. Transp. Syst., Funchal, Portugal, Sep. 19–22, 2010, pp. 812–818.

[53]. Federal Highway Administration (FHWA), Snow and ice, Washington, DC, USA, 2013 [Online]. www.ops.fhwa.dot.gov/weather/weather_events/snow_ice.htm

[54]. L. F. Musk, "Climate as a factor in the planning and design of new roads and motorways," in A. H. Perry and L. J. Symons (eds), Highway Meteorology. E. & F.N. Spon Press, London, 2003.

[55]. J. Collins, and M. Pietrzyk, "Wet and wild: Developing and evaluating an automated wet pavement motorist warning system," Transp. Res. Rec., J. Transp. Res. Board, vol. 1759, no. 1, pp. 19–27, 2001.

[56]. Federal Highway Admin. (FHWA), Best Practices for Road Weather Management, Washington, DC, USA, 2012.

[57]. J. Alcamo et al., Climate Change 2007: Impacts, Adaptation and Vulnerability. Contribution of Working Group II to the Fourth Assessment Report of the Intergovernmental Panel on Climate Change [M. L. Parry, O. F. Canziani, J. P. Palutikof, P. J. van der Linden, and C. E. Hanson, Eds]. Cambridge University Press, Cambridge, 2007, pp. 541–580.

[58]. V. Gornitz, S. Couch, and E. K. Hartig, "Impacts of sea level rise in the New York City metropolitan area," Global Planetary Changes, vol. 32, pp. 61–88, 2002.

[59]. C. Rosenzweig et al., "Developing coastal adaptation to climate change in the New York City infrastructure-shed: Process, approach, tools, and strategies," Climatic Change, vol. 106, pp. 93–127, 2011.

[60]. L. Goodwin and P. Pisano, "Current practices in transportation management during inclement weather," in Proc. Inst. Transp. Eng. Annu. Meet., Mitretek Systems ITS Division, Philadelphia, PA, USA, 2002.

[61]. Texas Dept. Transp., 2013—San Antonio District Flood Station Map and Information Webpage, Austin, TX, USA, 2013 [Online].

[62]. M. Nookala, and S. Bahler, Focus Group Evaluation of Advanced Traveler Information Systems in Minnesota. Minnesota Dept. Transp., St Paul, 2001.

[63]. L. Osborne, Jr., and M. Owens, Evaluation of the Operation and Demonstration Test of Short-Range Weather Forecasting Decision Support Within an Advanced Rural Traveler Information System. Univ. of North Dakota, Grand Forks, 2000 [Online]. www. itsdocs. FHWA.dot.gov/jpodocs_te/@9301!.pdf

[64]. A. H. Perry, and L. J. Symons, Highway Metrology. Univ. of Wales, Swansea, 1980. www.transguide.dot.state.tx.us/ITS_WEB/Frontend/default.html? r=SAT&p=San%20 Antonio&t=ess

[65]. J. Bogren, and T. Gustavsson, "RSI—Road status information A new method for detection of road conditions," in Proc. 17th Int. Road Weather Conf., Andorra, Portugal, 2014.

[66]. J. Norrman, M. Eriksson, and S. Lindqvist, "Relationships between road slipperiness, traffic accident risk and winter road maintenance activity," Climate Res., vol. 15, no. 3, pp. 185–193, 2000.

[67]. J. Bogren, T. Gustavsson, and H. Gaunt, Tema Vintermodell: VviSdata som underlag för beräkning av väglagsfördelning och olycksstatistik, Earth Sciences Centre, Göteborg University, Göteborg, Sweden, Rep. C75, 2006.

[68]. Federal Highway Admin. (FHWA), ITS Strategic Research Plan, 2010–2014, Progress Update 2012, Washington, DC, USA, 2012 [Online]. www.its.dot.gov/strategicplan/pdf/ ITS%20Strategic% 20Plan%20Update%202012.pdf

[69]. D. Veneziano, Z. Ye, and I. Turnbull, "Speed impacts of an icy curve warning system," IET Intell. Transp. Syst., vol. 8, no. 2, pp. 93–101, 2013.

[70]. "Severe weather warning system dual use safety technology (DUST) warning system," Phoenix, AZ, USA, 2013.

[71]. H. E. Brooks and C. A. D. III, "Normalized Damage from Major Tornados in United States 1980–1999," Weather Forecast., vol. 15, pp. 168–167, 2001.

[72]. B. Wolshon, E. Urbina, C. Wilmot, and M. Levitan, "Review of policies and practices for hurricane evacuation. I: Transportation planning, preparedness, and response," Nat. Hazards Rev., vol. 6, no. 3, pp. 129–142, 2005.

[73]. B. Barrett, B. Ran, and R. Pillai, "Developing a dynamic traffic management modeling framework for hurricane evacuation," Transp. Res. Rec., vol. 1733, pp. 115–121, 2000.

[74]. Florida DOT, "A new public safety focus for FDOT and ITS," Sunguide Disseminator Newsletter, Oct. 2011 [Online]. www.dot.state.fl.us/trafficoperations/ Newsletters/2011/2011–010-Oct.pdf#page=5

[75]. "Dust storm mitigation white paper by NMDOT," Santa Fe, NM, USA, 2013.

[76]. L. F. Musk, The Fog Hazard [A. H. Perry and L. J. Symons, Eds]. E. & F.N. Spon Press, London, 2003.

[77]. Federal Highway Admin. (FHWA), How Do Weather Events Impact Roads? Washington, DC, USA, 2013 [Online]. www. ops.fhwa.dot.gov/weather/q1_roadimpact.htm

[78]. "Highway fog warning system," Federal Highway Admin., Washington, DC, USA, 1999.

[79]. C. Schreiner, State of the Practice and Review of the Literature: Survey of Fog Countermeasures Planned or in Use by Other States. Virginia Tech Research Council, Arlington, VA, 2000.

[80]. M. A. Abdel-Aty, M. M. Ahmed, J. Lee, Q. Shi, and M. Abuzwidah, "Synthesis of visibility detection systems," Univ. Central Florida, Orlando, FL, USA, Final Rep., 2012.

[81]. "Future climate change," Washington, DC, USA, 2013.

[82]. F. H. Nofal, and A. A. W. Saeed, "Seasonal variation and weather effects on road traffic accidents in Riyadh City," Public Health, vol. 111, pp. 51–55, 1997.

[83]. R. Zimmerman, Global climate change and transportation infrastructure: Lessons from the New York Area. Federal Highway Admin. (FHWA), Washington, DC, USA [Online]. http://climate.volpe.dot. gov/workshop1002/zimmermanrch.pdf

[84]. NOAA. (1980). Impact assessment: U.S. social and economic effects of the great 1980 heat wave and drought, U.S. Department of Commerce, National Oceanic and Atmospheric Administration, Environmental Data and Information Service, Center for Environmental Assessment Services, Washington, DC, USA.

[85]. J. Isotalo, Seasonal Truck-Load Restrictions and Road Maintenance in Countries with Cold Climate, Transportation, Water and Urban Development Department, The World Bank, Washington, DC, USA.

[86]. D. Levinson et al., Cost/Benefit Study of: Spring Load Restrictions. Minnesota Dept. Transp., St Paul, MN, USA.

[87]. T. M. Hamil et al., "NOAA's future ensemble based hurricane forecast products," Bull. Am. Meteorol. Soc., vol. 93, no. 2, pp. 209–220, 2012.

[88]. B. Wolshon, E. U. Hamilton, M. Levitan, and C. Wilmot, "Review of policies and practices for hurricane evacuation II: Traffic operations, management, and control," ASCE Nat. Hazards Rev., vol. 6, no. 3, pp. 143–161, 2005.

[89]. P. J. Webster, G. J. Holland, J. A. Curry, and H. R. Chang, "Changes in tropical cyclone number, duration, and intensity in a warming environment," Science, vol. 309, no. 5742, pp. 1844–1846, 2005.

[90]. Alaska Department of Transportation (ADOT). (2013). Road weather information system (RWIS) public web site, Juneau, AK, USA [Online]. http://roadweather. alaska.gov

[91]. G. A. Lewis, E. Morris, S. Simanta, and L. Wrage, "Why standards are not enough to guarantee end-to-end interoperability," in Proc. 7th IEEE ICCBSS, Madrid, Spain, Feb. 25–29, 2008, pp. 164–173.

[92]. RITA, About ITS Standards, Jul. 5, 2014 [Online]. http:// www.standards.its.dot.gov/ LearnAboutStandards/StandardsDevelopment

[93]. M. Williams, D. Cornford, L. Bastin, R. Jones, and S. Parker, "Automatic processing, quality assurance and serving of real-time weather data," Comput. Geosci., vol. 37, pp. 353–362, 2011.

[94]. N. E. El Faouzi, R. Billot, P. Nurmi, and B. Nowotny, "Effects of adverse weather on traffic and safety: State-of-the-art and a European initiative," in Proc. 15th SIRWEC, Quebec City, QC, Canada, 2010, pp. 1–7.

[95]. Standing International Road Weather Commission (SIRWEC), What Is Sirwec? 2014. [Online]. www.sirwec.org/about.htm.

[96]. K. E. Trenberth, "Changes in precipitation with climate change," Climate Res., vol. 47, no. 1/2, pp. 123–138, 2011.

[97]. NRC (National Research Council) When Weather Matters: Science and Service to Meet Critical Societal Needs. National Academies Press, Washington, DC, 2010.

[98]. W. F. Krajewski, G. Villarini, and J. A. Smith, "Radar-rainfall uncertainties," Bull. Am. Meteorol. Soc., vol. 91, pp. 87–94, 2010.

[99]. H. Brodsky, and A. Hakkert, "Risk of a road accident in rainy weather," Accident Anal. Prevention, vol. 20, pp. 161–176, 1988.

[100]. K. Keay, and I. Simmonds, "The association of rainfall and other weather variables with road traffic volume in Melbourne, Australia," Accident Anal. Prevention, vol. 37, no. 1, pp. 109–124, 2005.

[101]. E. Hooper, L. Chapman, and A. Quinn, "The impact of precipitation on speed-flow relationships along a U.K. motorway corridor," Theor. Appl. Climatol., vol. 117, no. 1/2, pp. 303–316, 2014.

[102]. Y. Yang, H. Lu, Y. Yin, and H. Yang, "Optimization of variable speed limits for efficient, safe, and sustainable mobility," Transp. Res. Rec., J. Transp. Res. Board, vol. 2333, no. 1, pp. 37–45, 2013.

[103]. H. Lu, "Short-term traffic prediction using rainfall," Int. J. Signal Process. Syst., vol. 2, no. 1, pp. 70–73, 2014.

[104]. P. C. D. Milly et al., "Stationarity is dead: Whither water management?" Science, vol. 319, no. 30, pp. 573–574, 2008.

[105]. "Climate change 2007: Synthesis report—An assessment of the intergovernmental panel on climate change," Geneva, Switzerland, 2007.

[106]. United Nations, World Urbanization Prospects: The 2005 Revision, ESA/P/WP/200 2006.

[107]. "Associate program on flood management: Urban flood risk management—A tool for integrated flood management," Stockholm, Sweden, 2008.

[108]. K. C. Seto, and R. K. Kaufmann, "Urban growth in South China and impacts on local precipitation," in Proc. 5th Urban Res. Symp., Marseille, France, 2009.

[109]. J. H. Christensen, and O. B. Christensen, "A summary of the PRUDENCE model projections of changes in European climate by the end of this century," Climate Change, vol. 81, pp. 7–30, 2007.

[110]. J. H. Christensen, Ed., Climate Change 2007: The Physical Science Basis—Contribution of Working Group I to the Fourth Assessment Report of the Intergovernmental Panel on Climate Change. Cambridge Univ. Press, Cambridge, 2007.

[111]. C. R. Frei, S. Schöll, J. F. Schmidli, and P. L. Vidale, "Future change of precipitation extremes in Europe: Intercomparison of scenarios from regional climate models," J. Geophys. Res., vol. 111, no. D6, pp. D06105–1–D06105–22, 2006.

[112]. E. Buonomo, R. G. Jones, C. Huntingford, and J. Hannaford, "On the robustness of changes in extreme precipitation over Europe from two high resolution climate change simulations," Q. J. R. Meteorol. Soc., vol. 133, pp. 65–81, 2007.

9 Artificial Intelligence and Intelligent Transport Systems for Urban Transportation

Aiman Siraj, Vinod Kumar Shukla, Sonali Vyas and Soumi Dutta

CONTENTS

DOI: 10.1201/9781003218715-12

9.1 INTRODUCTION TO INTELLIGENT TRANSPORT SYSTEMS

Mobility has become a key concern in both the traditional and the modern world. Everyone has various reasons to travel or move from one place to another. It could be for small purposes like visiting a nearby store or transporting goods and services worldwide. Transportation has been made easier and advanced with time, with many changes being brought in by people's different means of transportation. Autonomous robots, like humans, even have the flexibility to form their selections and perform associate degree actions. They work as we do, as their systems are designed by implementing multiple techniques and algorithms that help them make decisions by predicting future outcomes. This ensures that there is little to no error in their work so that they can perform tasks efficiently and exceed their expectations. They can also be extremely useful in the transportation sector when used properly. Attacks or malicious practices on these robots might greatly harm human life. Automated vehicles also can cause a threat to the driver and the passengers if hacked. Hence appropriate and adequate security measures should be taken to maximize the secureness of these systems to prevent any mishaps that may occur in the future. In earlier times, people used to travel by foot or use animals to move goods or themselves. This was before the wheel was invented. The invention of the wheel gave rise to different ideas for creating more efficient modes of transportation. The Industrial Revolution in the 19th century also saw several inventions fundamentally change transport. It led to more competitive transportation, which significantly decreased the time spent going from one place to another.

9.1.1 MODES OF TRANSPORTATION

Modes of transportation can be classified into human-powered, animal-powered, land, water, air and others, such as

- Suborbital space flights;
- Pipelines used to transport liquid or gas through a system of pipes;
- Cable transport which involves balancing of load moving up and down and also a number of pulleys;
- Space flight for traveling to outer space.

The present-day means are more advanced and tend to be built to function automatically, which means they do not need human intervention or control. For example, there are self-driving cars, which are already programmed with their abilities and limitations to function automatically.

9.1.2 INTELLIGENT TRANSPORTATION SYSTEMS

Intelligent transportation systems (ITS), formerly called intelligent vehicle highway systems (IVHS), consists of communication technologies and information technologies that help maintain proper traffic and safety and improve driver's experiences as it reduces the risks of accidents [1].

Artificial intelligence (AI) is a term we have heard frequently in the past few years because of the rapid rate at which technology progresses. AI has already created a benchmark in terms of its achievements, capabilities and future scope. In the present era, AI is more accessible and in use than ever before. Subjects like machine learning, neural networks, evolutionary computation, robotics and speech processing are major AI subfields.

In other words, it can be said that it is a system in which communication and information technology is applied in the field of road transport, which also includes vehicles, users, infrastructure, and interfaces with other modes of transport. It includes CCTV surveillance systems, automated incident detecting systems, vehicle-to-roadside communication systems and driver information displays.

9.2 AI AND URBAN POPULATION

AI is a term we have heard frequently in the past few years because of the rapid rate at which technology is progressing. It is simply intelligence possessed by machines or more advanced robots. AI started small when it was just an idea, but now we can see AI almost everywhere [2–3].

Our smartphones are the most common example of AI that surely everyone already knows. Our smartphones contain AI software that runs the smartphone smoothly and optimizes it for better usability. AI in smartphones is only in its infancy and can only be seen in selected smartphones, but the scope for AI in smartphones is limitless. The first thing targeted by AI in smartphones was the camera. Using AI, the camera could better detect objects like a ball, cat or car, and using resources like the Internet could better process these images and develop sharper, crisp images. This is only one of the many ways AI is beginning to make simple tasks much easier. This is done to bring everyone on the same level. Using AI, even a novice could take good quality images without much knowledge of a smartphone camera. Another common example of smartphone AI is the ability to predict web searches and offer recommendations based on the person's viewing history or shopping history. These algorithms track our searches, learn about our habits, and then offer recommendations based on those searches or more information about that particular search. The machine learning algorithm can detect objects and faces in photos and videos and create search results based on them. For example, AI can detect a cake in an image and interpret it as a birthday celebration. So, the next time the user does not necessarily have to scroll down past hundreds of images; they can easily search for those images under the category birthday and give further recommendations for the system to learn.

Google defines AI as "intelligence demonstrated by machines in contrast to intelligence displayed by humans and animals." With the increasing problems in the world, the need for a reliable AI has become more important. In the urban population, where we have millions of people migrating from nearby towns and villages to try their luck in the big cities, there is a population explosion, which causes huge unemployment and terrible living conditions in some areas [4].

The data released by the United Nations estimates that the human population will reach 9.7 billion by the year 2050, and 70% of the population will be inhabited in cities with a population of around 10 million humans [5]. The rapid and extreme rise in city population threatens administration and management to prevent sanitation issues, mitigate traffic congestion and thwart crime. So, if many people live in the same city, the current cities and their services cannot successfully provide for the future challenges and the future population. This can be offered only through a smart city. Many components of a smart city come into play and co-exist and rely on each other for the perfect functioning of a smart city. Smart traffic management, smart parking, smart waste management and smart policing are some of the components of a smart city which make it "smart." Many or most of these problems can be solved using AI. Human progress throughout time has been slow-paced, but it is believed that AI can give us a boost [6].

The sharp rise in the graph is where AI can help humans progress. AI-enabled Internet of Things (IoT) uses advancements in the field of technology to create new and exciting experiences for its inhabitants of cities. That, in turn, was the foundation and created the concept of a smart city. A *smart city* can be defined as one that uses information and technology in different ways to make the lives of people living in urban areas easier without paying a price that is too high.

9.2.1 Examples

Following are the few applications of AI in the development of urban population, which are contributing a lot and also have a potential to grow further.

9.2.1.1 Waste Management

Waste management is essential for a thriving city or for any area for that matter. The requirement for a city's reliable and efficient waste management system is of profound importance. The increase in population increases the amount of waste produced and can harm society if left uncared for. An implication of Barcelona's waste management system shows how helpful technology can be in this aspect. The dustbins have sensors fitted into them, which alert the authorities by sending them notifications that the bin is about to be full. The authorities dispatch collection trucks that dispose of the waste away from the city and return the bins to their original location for the next cycle. It even has separate bins for the different types of waste, plastic, paper and glass. It also decreases the cost as the waste collection trucks do not have to keep going to every bin throughout the city to find out whether it is full. This saves fuel and also causes lesser pollution [7–8].

9.2.1.2 Crime Management

Crime is omnipresent, and in a large city, the need for smart policing is critical to keep the city safe for everyone. Wherever there is a massive influx of people in one area, unemployment increases because there aren't that many jobs being created as opposed to the number of people coming from other towns and villages to the city to seek employment opportunities. Thus, a few of these people, frustrated with life, resort to unfair means to sustain themselves and their families. Although this is wrong and because they did this just for food, a basic human necessity, crime is a crime, and that deed shouldn't go unpunished. A smart crime system will not stop crime from happening or alert the authorities before the crime has occurred, but it can help solve cases much faster. It monitors the movement and displays any unusual activity. This has already been implemented in Singapore. It enables the authorities to look into people smoking in no-smoking zones. Cities often have more skyscrapers than normal towns or villages; thus, it helps to monitor people living in skyscrapers who spread litter through their balconies. The cameras also enable monitoring of crowd density in an area and also the cleanliness in areas. As we know, during the recent pandemic, social distancing and hygiene these two have become the most important tools for fighting this pandemic. There is a huge need for cameras to monitor and supervise the number of people in areas to prevent the disease from spreading and infecting other innocent people. This, in turn, will help reduce crime rates in some of the most developed cities [9].

Every city needs the energy to function. Electricity is vital for our survival today. Without electricity, we will feel like cavemen of the early ages. The night sky in every city looks beautiful due to its lighting, and we need electricity to provide that lighting. Lighting is very important; it includes lighting at homes and offices and streetlights on roads and highways. Without streetlights, they can be many life-threatening accidents. These lights require a clean, renewable energy source to protect the environment. A lot of electricity is wasted if these lights are turned on or left on when they are not required. That wastage accumulated annually will be enough to provide electricity to a whole village, where most villages don't have access to a proper and uninterrupted electricity supply. The streetlights and lamp posts can be fitted with sensors that get activated when there is a human presence to allow the person to see the surroundings properly. The brightness of this light can be increased or decreased from a central unit which can work for many units simultaneously or individually. When not in use, the lamp can be triggered to provide minimum brightness so that people within range can see the light. The brightness of the light can then be increased as the person keeps coming closer to the lamp post. A bunch of lamp posts can be connected as a mesh network, where AI controls it. Using AI, a person can keep walking on the footpath where all the lights are dimmed. As a person comes close to a light, the brightness will keep increasing until they are completely perpendicular to the light, i.e., they are right below it, and the brightness will continue to decrease until it turns dim. The other street light senses the person and shows them the way. This not only saves much energy, at the same time, is very futuristic, but it can also prevent many crimes. People are often afraid to go out because of the dark or are afraid of thefts that may occur in the dark. If a person is following them, the lights sense that person and will not become dim, alerting them that someone is following them. The person can either inform the police or become more careful and alert. This can help in reducing many crimes in a way.

9.2.1.3 Housing

Housing becomes a big part of how successful the smart city is, and considering that most of these cities are metropolitans, it becomes even more important. Buildings consume much electricity and require better security than most other houses because many people live together in a building [10]. A compromise on security on one's part can lead to a big problem in the future for everyone included in the building [11]. Everyone in a building may have different requirements of temperatures, and to solve that, AI uses a machine-learning algorithm to maintain a constant temperature throughout the building and not varying temperatures in different parts of the building. AI can also control the lighting of the different sections of the buildings. Since lighting already consumes much electricity, when there is no one in the corridors or the lifts or the common area, the lights controlled by an algorithm are turned off. Through the sensors installed, they sense when there is a person in the corridor, and they light up. This is an efficient way in which it caters to the need of the people, and at the same time, it helps in conserving electricity. Since AI is based on a machine-learning algorithm and can accumulate and retain knowledge, every sensor, camera and monitor installed in the building can be integrated into the main management system, which monitors and captures every movement in the building. Due to it being able to learn, it can also modify the settings and continue to self-tune itself. That way, it keeps updating itself without the need to do it physically, and it also fixes bugs on its own rather than having a developer fix them. Smart buildings are a thing of tomorrow and are very important because, in the future, buildings are going to be assembled and not constructed. These buildings are designed to be easily modified, allowing quicker re-integration of the engineering systems, which saves much time for the builders as they do not have to install the smart software into the building again and again to begin the process of the system's learning. The same old system, which had already, with the advancement in time, learned a lot about the building and tuned itself to its residents, can be slightly modified to be in tune with the modifications in the building, and then it is all good to go [12–13].

9.2.1.4 Farming

Since farming has existed since civilization began, it is evident that a smart city should also have an innovative and efficient farming system to feed the ever-growing population. Traditional farming methods involve using older equipment and farming techniques which may or may not result in under-utilization of resources. In order to avoid wastage of resources, the use of AI can also be beneficial in this sector because smart farming is software-managed and sensor-monitored. Smart farming can revolutionize the agricultural sector using the cloud and AI for tracking, monitoring, automating and analyzing agricultural operations. Smart farming uses resources like sensors for soil scanning, which provides data like light, humidity and temperature. These statistics come into play later when the optimum amount of light, humidity, water or temperature is required for a particular kind of crop which increases its quality and quantity. Remote monitoring of the fields is also possible via satellites and drones, which keep a complete 24-hour vigilance over the field, better than employing people to watch over the fields. AI decides according to wind speed, soil moisture, sunlight, temperature, humidity, and fertilizer according to the data received from the sensors. Then it decides which crop is most suitable for that particular patch of land. This, in turn, increases crop productivity and prevents soil misuse. Even when it is time to harvest the crop, AI-enabled machines can do that much faster and more efficiently than we humans can do. For example, while harvesting sugarcane, the cane has to be cut at a certain height, and the remaining part of the crop needs to be planted back for the next harvest. Combine harvesting machines enabled with AI can extract the sugarcane crop, cut it to a fixed height and plant the remaining crop back into the soil, all in one go. For a machine to do it, it is done faster with less or no wastage of any resources and uniform throughout [14–15].

9.2.1.5 Manufacturing

The manufacturing industry keeps undergoing massive changes every few decades, and with these changes come advancements in technology, making manufacturing more accessible and efficient. AI in manufacturing will come as a revolution to the whole manufacturing industry. AI brings with it a fully integrated, collaborative manufacturing system capable of responding in real time to the needs and demands of the customers. Introducing AI in the manufacturing industry merges virtual and physical systems and can lead to many new innovations. The AI-controlled system can optimize the number of resources on the plant floor leading to higher quality products, higher efficiency, improved productivity and lower chances of any mishap occurring. In the automobile industry, machines fitted with sensors can operate the complete assembly line from the frame to a fully functional car. These machines, fitted with sensors and AI algorithms, can deal with the most intricate details of the vehicle and have a precision of about one-tenth of a millimeter, leaving no room for error.

9.3 AI IN TRANSPORTATION

In the present era, AI is more accessible and in use than ever before. Subjects like machine learning, neural networks, evolutionary computation, robotics and speech processing are major sub-fields of AI. Even the transportation industry is embracing artificial intelligence for various reasons, mainly to be more systemized. Machine learning and AI technology are now taking up the role of a crew member in aircraft. They adjust the aircraft's control surfaces by just using the sensor data based on flight conditions.

9.3.1 EXAMPLES OF AI IN TRANSPORTATION

Following are the few examples based on land, air and water.

9.3.1.1 On Land

AI is being used in various ways within transportation it becomes more advanced as the industry evolves towards automation.

9.3.1.1.1 Autonomous or Self-Driving Vehicles

This is one of the most exciting modern world innovations, which has become a part of reality. In autonomous vehicles, data is transmitted in real time. Though it is a boon for the industry, sometimes minor destruction of the processes can lead to a catastrophe, which is dangerous for the driver and other vehicles on the road [16]. Some major companies that have developed autonomous vehicles are Mercedes-Benz, Toyota, Audi and Volvo. Tesla Autopilot is a suite of advanced driver assistance systems with features like lane centering, self-parking and semi-autonomous navigation. But the driver is responsible for the functioning of these features and the car needs constant supervision.

9.3.1.1.2 Traffic Management Systems and Law Enforcement

Again, due to its ability to process control and optimize, AI can be used to manage traffic in order to prevent congestion. This feature can have a lot of benefits in countries like India, where there are no proper road systems, creating lot of traffic jams and loss of time.

9.3.1.1.3 Smart Roads

Roads provide millions of opportunities that people have just started to explore. It also allows us to rethink what changes can be made on the roads, slowly and gradually, through experimentation, to provide a safe experience for the users [17]. Intelligent planning and thinking can help build less complicated road networks and equip them with the latest and greatest IoT and communication technologies. These changes are taking place on a greater scale in developing countries than in developed cities and countries. Despite many technological advances made to automobiles, comparatively, there are very few changes made to asphalt roads. A few future features we might see on the roads include glow-in-the-dark road markings, which are useful at night, wind-powered lights, electric priority lanes in which electronic vehicles can charge on the go, and solar roadways. In a way, it helps in making use of solar energy. Other smart road technologies are license plate recognition, automatic toll collection, smart roadside digital signage, bridge, tunnel condition monitoring, and security.

9.3.1.2　On Aviation and Maritime

The global aviation ITS market has provided a report on the different systems and their application areas. It provided insights on the system market for tracking and monitoring, smart gate, self-service baggage market, kiosks and information displays. Application areas include security and surveillance, shuttle bus tracking, and aircraft and emergency management. In the global maritime ITS market, areas like automatic long-range identification, traffic management, navigation and information services have been included [18].

9.3.2　Traffic Management and Parking Control Using AI

Traffic management using AI is a breeze. It needs no or minimal human intervention, and as is already known, AI can learn over time and also the ability to retain that knowledge. Managing traffic using AI involves using road-surface sensors and CCTV cameras to send real-time updates to the central traffic management system. No longer has a person had to physically assess the movement of traffic and then send relative information to the traffic management system. The data feed in the system received from the road camera can easily detect traffic congestion. Based on that, it notifies users traveling on that road about congestion, traffic signal malfunctions and possible time delays in reaching their destination. Also, the vehicle license plates can be known using the camera systems, and details about the vehicle and the owner can be easily obtained if needed. For example, if a vehicle is stolen and the owner lodges a complaint with the local police enforcement, then using the help of the traffic management system, the police can get to know the location of the vehicle, and the owner can get their stolen vehicle back much faster. Also, the police can get information about the owner's license status, the vehicle's insurance status, and whether the vehicle is serviced and fit for the road. Cameras installed in trains, buses and metros can also monitor the movement of people. For example, in light of the recent COVID-19 pandemic, wearing face masks and gloves and maintaining a safe distance from other people is very important. Cameras can alert if anyone on any of the modes of transport, train, bus or metro is not wearing their face masks or gloves or maintaining social distancing. The person can be identified and fined or penalized by the system. Nowadays, cars have abundant AI-reliant features like lane-keep assist, lane-departure warning, cruise control, seatbelt warning, collision warning and parking assist [19–20].

In cities, another major problem apart from traffic is parking, and it sometimes takes a long time for people to find suitable parking. Most of the time, they have to park far away from where they need to be and walk to and fro to reach their cars; sometimes, they don't even find parking. AI has a solution to this parking hassle. Using the road sensors installed on the roads, AI can effectively monitor which parking space is empty or occupied and create a real-time parking map. This saves much time for the people who struggle finding parking spaces and also helps reduce pollution caused by the fuel wasted while searching for the parking spot.

9.4 AUTONOMOUS ROBOTS AND THEIR ISSUES

Autonomous robots, like humans, even have the flexibility to form their selections and perform associate degree actions. A very autonomous robot may understand its surroundings, build selections supported by what it perceives and be programmed to acknowledge and actuate a movement or manipulation among that surroundings. With relevancy quality, as an example, these decision-based actions, however, are not restricted to the subsequent basics: beginning, stopping, and maneuvering around obstacles.

- Autonomous robots utilize infrared or ultrasound sensors to check obstructions, allowing them to explore the impediments without human management. Additionally, developed robots utilize audio system vision to check their surroundings; cameras offer them profundity discernment, and programming enables them to seek out and increasingly prepare objects.
- Autonomous robots are helpful in occupied things, like a medical clinic. Instead of representatives going to their posts, a self-sustaining robot will quickly convey research lab results and patient examples.

9.4.1 EXAMPLES OF AUTONOMOUS ROBOTS

Following are different examples of autonomous robots.

9.4.1.1 Roomba

The Roomba is a series of autonomous robotic vacuum cleaners by iRobot. Roomba options a group of sensors that alter it to navigate the ground space of a home and clean it. For example, Roomba's sensors will notice the presence of obstacles and dirty spots on the ground and sense steep drops to keep it from falling down the steps. Roomba uses two severally operative facet wheels that permit 360-degree turns in situ. A rotating, three-pronged spinner brush sweeps debris from square corners to the cleansing head [21].

9.4.1.2 Connie, Hilton's Robot Concierge

Connie is a robot caretaker employed by Hilton. With its speech recognition capabilities, the robot uses an artificial intelligence platform developed by IBM and is ready to act with guests and reply to their queries. The system additionally learns and adapts with every interaction, raising the answers it provides [22].

9.4.1.3 Amy Waitress

The robot waiter is an associate innovative assistant for any service sector. The robot waiter will perform reception duties, deliver food and drink, generate interactive expertise and supply informatory and informative explanations. It may be employed in hotels for space service, in automobile dealerships, restaurants and cafes [23–24].

9.4.1.4 Pneuborn

Pneuborn is a robot that behaves like an infant. It was developed to study the musculoskeletal systems in babies. It will teach itself to crawl, sit, and stand. Osaka University had developed it [25].

9.4.2 SECTORS IN WHICH THE DEMAND FOR SUCH ROBOTS WILL INCREASE

In the time to come, the demand for robots will be very high. Moreover, robots will be demanded in all fields. The following are a few examples [26].

9.4.2.1 Medical and Healthcare

Robots have already been aiding surgeons around the globe. In line with BIS analysis, it has been known that surgical systems retain the largest share of the medical artificial intelligence market. Different kinds of robots are used for prescription drug testing, handling, and disinfecting hospitals. Surgical robots have the maximum accuracy for movements, reduction in potential errors, reduction in recovery times and fewer risks of complications.

9.4.2.2 Agriculture

Automation has been an essential part of agriculture for many years. The number of agricultural robots shipping will reach 727,000 units annually by 2025, predicts Tractica analysis. Robot area units are already helping milk cows and plant and manage crops. Even the Internet of Things (IoT) is employed in factories and warehouses, and robots for autonomous vehicles like tractors.

9.4.2.3 Emergency and First Response

Mobile robots are used as first responders for any damages after a hurricane or tornado. For example, a drone can spot survivors after an earthquake and navigate beneath a collapsed structure.

9.5 BENEFITS OF ITS

Countries seek to harness modern technologies and innovations to transform their transportation and industrial sectors, which in a way also helps them increase the amount of profit they can earn through tourism or the regular public.

With the increasing number of vehicles on the road in both developed and developing countries, there is a high risk of road accidents which is a significant public safety problem. It is also because people tend to ignore traffic rules while on the road, either as pedestrians or drivers. Hence, deploying ITS will alter vehicles to communicate autonomously with different vehicles around them and margin infrastructures. It can open the door for a wide range of road safety and helpful driver applications.

According to the United Nations, it is estimated that about 1.3 million people worldwide are killed in road accidents every year. Most traffic fatalities occur on rural or underdeveloped road networks. This number could reach about 1.9 million if left unchecked [27]. This would also create much economic loss for the country. According to this report, those in the age group 5–29 are more vulnerable to

TABLE 9.1

Road Fatality Data for 2020 Compared to the Average 2017–2019 [28]

Country	2020 road deaths	2017–2019 road deaths
Argentina	3322	5338
Australia	1108	1182
Austria	344	413
Belgium	499	619
Canada	1747	1854
Chile	1794	1951
Colombia	5447	6570

accidents, and males who are younger than 25 years are found to be more vulnerable to accidents. Table 9.1 shows fatalities by country.

Though connecting vehicles in a wireless way will prove to be beneficial, it can also give rise to security threats, which can make the applications and the driver more vulnerable to accidents. Therefore, it is mandatory to fully analyze and test an intelligent transport system before implementing it in the real world.

9.5.1   How Does ITS Work?

Various factors contribute in order for ITS to work properly. The following are the main important factors for the operation of ITS [1].

9.5.1.1   ITS Combining Technologies

The technical core of ITS is applying knowledge and management technologies to transit operations. These technologies embrace communications, automatic management, computer hardware and software. It requires proper knowledge from many backgrounds, including most engineering backgrounds like civil engineering, mechanical engineering, electrical engineering and studies related to the industry. The lack of timely and accurate information and non-cooperation from individuals are the major causes of problems in transportation.

9.5.1.2   ITS Enabling Technologies

Few technologies that help function intelligent transport systems are GPS, laser sensors, fiber-optics, digital maps, compasses and display technologies.

These technologies can be classified into the following.

9.5.1.2.1   Data Acquisition

Radar and ultrasonic traffic sensors and closed-circuit television (CCTV) provide live images of traffic to monitor it and serve as evidence in case any accidents occur.

9.5.1.2.2   Data Processing

This is done by using the data fusion process. A *data fusion process* is where multiple data sources are integrated to produce more efficient, useful and consistent data compared to the data provided by an individual source. Automatic incident detection (AID) might use data processing.

9.5.1.2.3 *Data Communication and Distribution*

It helps convey messages in commercial vehicle operations (CVO), electronic toll collection (ETC), or to collect travelers' information. Data distribution consists of many devices such as car radio, television, cellular telephones or handheld devices.

9.5.2 HYPERLOOP: AN INNOVATION FOR GLOBAL TRANSPORTATION

Hyperloop technology is an example of an ITS in smart cities. It is being built for transportation for people and goods across land and ocean in much less time than what we have now in the present world. This technology can have a massive impact on environment-friendly transportation in the future. It also contributes to the growth of the economy. According to various research, it is said that the hyperloop can be the next sustainable mode of transport. But it might not be considered a global solution for the challenges in transportation; rather, it is just a new and distinct type of vehicle that is much more efficient than what we have in the present day. As this technology is fully electronic, it can be considered a solution to reduce the environmental impacts of mobility.

9.5.3 BENEFITS OF HYPERLOOP IN THE POST-COVID WORLD

The public transport sector has been hit the hardest since the breakout of the COVID-19 virus around the world. Public transportation systems are usually the main area where most people come together and travel to their respective destinations. This increases the chance of spreading the virus even if there is just one person who may be infected. Due to contamination concerns, people might also try their best not to use public transport as much as possible, which is another major blow for the transportation sector.

So, in the hyperloop, pods (or compartments) can be customized according to the number of people and whether they need space for hauling cargo or not. It also reduces the need for human contact as everything is automated. It will also comprise the role of biometrics. It allows a person to call a pod, get to the station and be scanned for luggage and temperature. Though the hyperloop is not yet in operation, it is not a work of fiction. The first hyperloop is scheduled to be launched in 2030. Soon, it will be the fifth mode of transport operating in smart cities [29].

9.6 COMMON CYBER SECURITY PROBLEMS WITH AI SYSTEMS AND OTHER ISSUES

9.6.1 UNPRECEDENTED ATTACKS

The amount of valuable info that resides on multiple information sources has matured exponentially from the first days of the laptop. The chance for organizations of all sizes to have their information compromised grows because the range of devices that store confidential information increases.

9.6.2 CYBER UNDERCOVER WORK

Both large and small organizations are setting out to store a minimum of a number of their information within the cloud. Right scale recently found that personal cloud adoption augmented to 77 among organizations; hybrid cloud computing augmented further.

9.6.3 DATA STEALING

Data stealing is the act of stealing digital info held on computers, servers or electronic devices of associate degree from unknown victims with the intent to compromise privacy or acquire control. Information will vary from monetary information, like debit card numbers or bank accounts, to non-public info, like Social Security numbers, driver's license numbers, and health records.

DDoS Attacks (overloading the server beyond its capacity), malware (providing unauthorized access), phishing scams (pretending to be legitimate user/service providers), internal misuse (by internal employees or stakeholders) and other similar attacks should also address while discussing the cyber security problem with AI.

The Cyber Intelligence Sharing and Protection Act (CISPA H.R. 3523 (112th Congress), H.R. 624 (113th Congress), H.R. 234 (114th Congress)) is a projected law within the United States that might allow the sharing of net traffic info between the US government and technology and producing corporations. The bill's declared aim is to assist the US government in investigating cyber threats and ensure networks' security against cyberattacks. The legislation was introduced on November 30, 2011, by Representative Mike Rogers (R-MI) and 111 co-sponsors [30].

In general, cyber security for robots and autonomous systems is set to suffer similar cybersecurity issues that computers have faced for many years. This cannot be solely worrying for vital tasks like those performed by surgical or military robots; however, conjointly for home robots like vacuum cleaners or teleconferencing robots compromise the privacy and safety of their house owners. In order to supply the simplest performance, they are ceaselessly gathering data. Beneath these circumstances, if these robot's security is compromised, then a two-dimensional security downside arises: first, security problems relating to the virtual facet of the automaton (data, communications and so on); and second, those issues related to a physical facet that issues each automaton and user integrity. The state of the art presents the "cyber-physical security" term to cover virtual and physical issues.

9.7 IMPACT OF CYBERSECURITY ISSUES

Robots, as they exist currently, are typically massive significant machines that are preponderantly employed in a production setting as machine-driven vehicles. As the use of robots becomes much thought, impact on human safety should be considered. The impact of a cyber-attack on a military robot or drone is the most feared since they are supposed to conduct police investigation or cause irrecoverable damages, which may end in injuring innocent civilian or military lives. Human safety has become a region of concern about the utilization of machine-controlled vehicles.

Even a tiny error may risk the life of the driving force and the passengers; a few cases have already been according to such accidents [31].

Future machine-driven vehicles will be driverless, and there could also be no choice but to override the vehicle by the traveler. They even are not passengers and are solely used for transporting the product. Eldercare robots will be designed to cohabitate with the humans they assist or look after. Therefore, it is simple to assume that an associate in nursing eldercare automaton would solely affect the protection of one or two people.

9.7.1 COMMON SOLUTIONS TO CYBERSECURITY ISSUES

The following are a few easy-to-implement solutions related to cybersecurity.

9.7.1.1 Require Employees to Use Strong Passwords and Alter Them Periodically

Creating a robust, secure laptop is probably the most effective issue to strengthen the protection of your system. A user has to create a password that uses special characters, letters and numbers.

9.7.1.2 Use the Internet with Precaution

It is a necessary to grasp that even the safest websites can contain spyware and malware. All it takes is one click for the laptop to be infected with spyware. There are many websites that seem to operate like real websites. Examine the name of a website and guarantee its veracity.

9.7.1.3 Avoid Spam Email

Users of email need to look out once they are victimized. Hackers can exploit their email in many ways, including activity viruses in attachments. It is vital not to open or scan unknown email.

9.7.1.4 Scheduled Update of the System

It is vital to run regular system updates. Otherwise, the bugs in your system might be exploited by hackers.

9.7.1.5 Use of Firewall

A network firewall may be a very important tool; it acts as protection around your laptop and blocks unauthorized access.

9.8 FUTURE SCOPE OF ITS

When discussing future possibilities, one thing is clear: it's not just about collecting data. It is beyond that; as we process data, the machines can make their own decisions by predicting what will happen next. These can be converted to a real-time system which relays information about each mode of transportation through different delivery sources [32]. This data can be reflected on the side of the road or through an application or web page to convey the important happenings to the drivers on the road. This data can also be used to implement several strategies to overcome road issues so people can drive safely. This will not only ease traffic problems but will give a head start to people who are in a hurry to take less congested routes. It will also drastically reduce the number of accidents. In case of emergencies, ambulances and police vehicles can reach their destinations faster and decrease response time which will, in turn, save more lives. It also improves the economic condition of a country by creating more jobs and employment opportunities. This particular system has a huge future scope which can be achieved by persistence in the coming years.

9.9 CONCLUSION

The chapter discusses ITS and AI's role in this system. The way AI can develop the urban population is also briefly explained. Several ways to manage traffic and parking issues using AI are addressed. Several cybersecurity issues and methods to overcome them are mentioned. The autonomous robot is a mechanism that performs tasks with a high degree of autonomy. It is thought of as a subfield of AI, robotics and data engineering. Examples of autonomous robots (in current use) are the Roomba, a robot vacuum cleaner; the Amy waitress dining service robot used in restaurants to attract customers; Connie, the caretaker who receives the customer at the Hilton and also has artificial speaking intelligence to interact verbally with hotel guests; and Pneuborn, a robot with the characteristics of a human infant developed for the study of an infant's musculoskeletal system. These autonomous robots are used in the field of medical science to perform surgery or to treat patients. They are also used in agriculture to manage growing crops and automatically harvest them according to the season. They are also designed to help during human-made or natural disasters. Automated vehicles also can cause a threat to the driver and the passengers if hacked. The future scope of ITS is also mentioned. AI is still a new concept, and mistakes are bound to be made along the way while we try to improve it. AI is still only functioning in a few sectors; it has to spread to the other sectors to create an integrated system where everything is AI enabled. The pandemic has reminded us how mostly everything has a remote solution, it just needs to be found, and AI is the way to go in this case. AI's relative newness should not be confused with an inability to perform. The scope for ITS is abundant and will play a huge role in the future.

REFERENCES

[1]. Intelligent Transportation Systems [online] www.see-industry.com/en/intelligent-transportation-systems/2/930/

[2]. D. Mathew, V.K. Shukla, A. Chaubey and S. Dutta, "Artificial Intelligence: Hope for Future or Hype by Intellectuals?," in 2021 9th International Conference on Reliability, Infocom Technologies and Optimization (Trends and Future Directions) (ICRITO), IEEE, Noida, India, 2021, pp. 1–6, doi:10.1109/ICRITO51393.2021.9596410.

[3]. F.F. Petiwala, V.K. Shukla and S. Vyas, "IBM Watson: Redefining Artificial Intelligence Through Cognitive Computing," in M. Prateek, T.P. Singh, T. Choudhury, H.M. Pandey and N. Gia (eds), Proceedings of International Conference on Machine Intelligence and Data Science Applications. Algorithms for Intelligent Systems. Springer, Singapore, 2021, doi:10.1007/978-981-33-4087-9_15

[4]. Accueil Wathinote initiative IA Artificial Intelligence Definition, Ethics and Standards, The British University in Egypt, 2019 [Online]. www.wathi.org/artificial-intelligence-policies-in-africa-over-the-next-five-years-researchgate-2019/

[5]. United Nations, Department of Economic and Social Affairs [Online]. www.un.org/en/desa/population-growth-opportunity-action-sdgs-climate-un-desa-report-says

[6]. S. Hansain, D. Gaur and V.K. Shukla, "Impact of Emerging Technologies on Future Mobility in Smart Cities by 2030," in 2021 9th International Conference on Reliability, Infocom Technologies and Optimization (Trends and Future Directions) (ICRITO), 2021, pp. 1–8, doi:10.1109/ICRITO51393.2021.9596095.

[7]. M. Goel, A. H. Goyal, P. Dhiman, V. Deep, P. Sharma and V.K. Shukla, "Smart Garbage Segregator and IoT Based Waste Collection system," in 2021 International Conference on Advance Computing and Innovative Technologies in Engineering (ICACITE), 2021, pp. 149–153, doi:10.1109/ICACITE51222.2021.9404692.

[8]. R.M. Ibrahim, V.K. Shukla, A. Yadav, S.R. Pillai and N. Pandey, "Electronic Waste Mitigation using Photovoltaic Systems," in 2021 9th International Conference on Reliability, Infocom Technologies and Optimization (Trends and Future Directions) (ICRITO), 2021, pp. 1–4, doi:10.1109/ICRITO51393.2021.9596361.

[9]. N. Nazar, V.K. Shukla, G. Kaur and N. Pandey, "Integrating Web Server Log Forensics through Deep Learning," in 2021 9th International Conference on Reliability, Infocom Technologies and Optimization (Trends and Future Directions) (ICRITO), 2021, pp. 1–6, doi:10.1109/ICRITO51393.2021.9596324.

[10]. R. Khan, V.K. Shukla, B. Singh, S. Vyas, "Mitigating Security Challenges in Smart Home Management Through Smart Lock," in T.P. Singh, R. Tomar, T. Choudhury, T. Perumal and H.F. Mahdi (eds), Data Driven Approach Towards Disruptive Technologies. Studies in Autonomic, Data-driven and Industrial Computing. Springer, Singapore, 2021, doi:10.1007/978-981-15-9873-9_7

[11]. S. Ibrahim, V.K. Shukla and R. Bathla, "Security Enhancement in Smart Home Management through Multimodal Biometric and Passcode," 2020 International Conference on Intelligent Engineering and Management (ICIEM), 2020, pp. 420–424, doi:10.1109/ICIEM48762.2020.9160331.

[12]. V.K. Shukla and B. Singh, "Conceptual Framework of Smart Device for Smart Home Management Based on RFID and IoT," in 2019 Amity International Conference on Artificial Intelligence (AICAI), 2019, pp. 787–791, doi:10.1109/AICAI.2019.8701301.

[13]. S.F.N. Zaidi, V.K. Shukla, V.P. Mishra, B. Singh, "Redefining Home Automation Through Voice Recognition System," in A.E. Hassanien, S. Bhattacharyya, S. Chakrabati, A. Bhattacharya and S. Dutta (eds), Emerging Technologies in Data Mining and Information Security. Advances in Intelligent Systems and Computing, vol. 1300. Springer, Singapore, 2021, doi:10.1007/978-981-33-4367-2_16

[14]. S. Vinod Kumar, R.S. Nair, and F. Khan, "Smart Irrigation-Based Behavioral Study of Moringa Plant for Growth Monitoring in Subtropical Desert Climatic Condition," in AI, Edge and IoT-based Smart Agriculture. Academic Press, Cambridge, 2022, pp. 227–240.

[15]. S. Murlidharan, V.K. Shukla and A. Chaubey, "Application of Machine Learning in Precision Agriculture using IoT," in 2021 2nd International Conference on Intelligent Engineering and Management (ICIEM), 2021, pp. 34–39, doi:10.1109/ICIEM51511.2021.9445312.

[16]. "Autonomous System (AS)" [online]. www.techopedia.com/definition/11063/autonomous-system-as

[17]. "Transportation Case Study" [online]. www.lanner-america.com/solutions/transportation/smart-roads/

[18]. S.H. Siddiqi, V.K. Shukla and R. Bathla, "Redefining Efficiency of TCAS for Improved Sight through Image Processing," in 2020 Research, Innovation, Knowledge Management and Technology Application for Business Sustainability (INBUSH), 2020, pp. 151–155, doi:10.1109/INBUSH46973.2020.9392159.

[19]. A. Siraj and V.K. Shukla, "Framework for Personalized Car Parking System Using Proximity Sensor," in 2020 8th International Conference on Reliability, Infocom Technologies and Optimization (Trends and Future Directions) (ICRITO), 2020, pp. 198–202, doi:10.1109/ICRITO48877.2020.9197853.

[20]. F.F. Petiwala, V.K. Shukla, V.P. Mishra and S. Saini, "Smart Parking System through Automation in License Plate Recognition," in 2021 9th International Conference on Reliability, Infocom Technologies and Optimization (Trends and Future Directions) (ICRITO), 2021, pp. 1–7, doi:10.1109/ICRITO51393.2021.9596554.

[21]. "Roomba" [online]. https://en.wikipedia.org/wiki/Roomba

[22]. "8 Examples of Robots Being Used in the Hospitality Industry" [online]. www.revfine.com/robots-hospitality-industry/

[23]. U.S. Mohammed, V.K. Shukla, R. Sharma, A. Verma, "Next Step to the Future of Restaurants through Artificial Intelligence and Facial Recognition," in J.M.R.S. Tavares, P. Dutta, S. Dutta, D. Samanta (eds), Cyber Intelligence and Information Retrieval. Lecture Notes in Networks and Systems, vol. 291. Springer, Singapore, 2022, doi:10.1007/978-981-16-4284-5_9

[24]. "Amy Waitress" [online]. www.servicerobots.com/amy-waitress/

[25]. "Pneuborn" [online]. https://robots.ieee.org/robots/pneuborn/#:~:text=Pneuborn%20is%20a%20robotic%20infant,pattern%20generators%20and%20learning%20algorithms.

[26]. Kayla Matthews, "6 Industries Where Demand for Robotics Developers will Grow by 2025" [online]. www.therobotreport.com/6-industries-demand-robotics-developers-grow-2025/

[27]. A.L. Madana and V. K. Shukla, "Conformity of Accident Detection Using Drones and Vibration Sensor," in 2020 8th International Conference on Reliability, Infocom Technologies and Optimization (Trends and Future Directions) (ICRITO), 2020, pp. 192–197, doi:10.1109/ICRITO48877.2020.9197783.

[28]. International Transport Forum, Road Safety Annual Report 2021. www.itf-oecd.org/sites/default/files/docs/irtad-road-safety-annual-report-2021.pdf, Accessed on 16.10.2022

[29]. Jennifer Bell, "Expo 2020 Dubai: Virgin Hyperloop to Unveil New High-Speed Passenger Pods," Al Arabiya English. https://english.alarabiya.net/News/gulf/2021/09/27/Expo-2020-Dubai-Virgin-Hyperloop-to-unveil-new-high-speed-passenger-pods, Accessed on 15.10.2022

[30]. Congress.gov, "H.R.624 — Cyber Intelligence Sharing and Protection Act 113th Congress (2013–2014)," www.congress.gov/bill/113th-congress/house-bill/624, Accessed on 15.10.2022

[31]. G.W. Clark, M.V. Doran and T.R. Andel, "Cybersecurity Issues in Robotics," in 2017 IEEE Conference on Cognitive and Computational Aspects of Situation Management (CogSIMA). IEEE, Piscataway, 2017, pp. 1–5.

[32]. V.K. Shukla, L. Wanganoo and N. Tiwari, "Real-Time Alert System for Delivery Operators through Artificial Intelligence in Last-Mile Delivery," in L. Garg, C. Chakraborty, S. Mahmoudi, V.S. Sohmen (eds), Healthcare Informatics for Fighting COVID-19 and Future Epidemics. EAI/Springer Innovations in Communication and Computing. Springer, Cham, 2022, doi:10.1007/978-3-030-72752-9_20

10 Intelligent Communication Systems for Urban Things

Yashi Yashi, Rahul Nijhawan and Deepshikha Bhargava

CONTENTS

10.1 INTRODUCTION

In a couple of years, we have seen an abrupt expansion in human population. The possibility of smart urban areas has been set up to make human existence more satisfied and more secure. With time, the portrayal of the brilliant city has additionally been advanced [1]. As per a few creators, smart urban areas are an amalgamation of mind-boggling frameworks (or actual capital) with data, correspondence, and social foundation availability and quality. Assuming that savvy urban communities are

DOI: 10.1201/9781003218715-13

set up with the appropriate organization and innovation, they will work on human life's nature. By joining the capacities of the actual world with the comprehension of the computational world, the Internet of Things (IoT) assumes a basic part in savvy city advancement. The objective is to work on the common sense of capacities in reality by expanding the capabilities of things in the genuine city. With cell phones, PCs, and sensors contracting, interest in correspondence innovations that give blunder-free networks has filled drastically in smart urban areas course (ICT). The smart city's construction and upkeep will necessitate cutting-edge integrated technologies. Sensors, electronics, and networks utilize computerized systems with databases, decision-making algorithms, and tracking.

Smart cities are fully reliant on network connectivity. It necessitates great speed, dependability, availability and features in today's networks. Modern cities face numerous complex issues as they endeavor to keep up with the speed of constant development. To address this issue, organizational adjustments are required, typically prioritizing cutting-edge technologies and Internet connectivity. IoT can be leveraged to provide better value and services by exchanging data with manufacturers, operators and other connected devices. Connecting real and virtual worlds with a significant number of electronics spreads in many areas, such as households, automobiles, streets, buildings, and many other surroundings. It is regarded as a key research and development concept.

The smart city models are created with real-time functionality in mind. It's done using sensed data collected at regular intervals and the usage of intelligent laboratories to make monitoring and design easier. With time, new ICT emerges. It is intrinsically networked and permits extensive connections across multiple disciplines. Citizens must be able to mix their knowledge with specialists tasked with developing these technologies, which must be a part of the management process. This raises concerns about privacy and security, which are critical to the effective operation of smart cities. Information security is required for a smart city's economic advancement to be successful. A smart city's essential parts are a smart city's smart economy, climate, government, individuals, versatility, and a brilliant way of life. Urban areas can profit from utilizing data and correspondence advances (ICTs) with naturally maintainable and monetarily possible arrangements. Upgrades could be made, for instance, by efficient water the board dependent on ongoing data trade, air quality, electromagnetic field observing, and public transportation frameworks organized utilizing satellite information. Urban communities started to utilize ICT to offer further developed types of assistance and personal satisfaction to occupants some time ago.

10.1.1 Application Domains of Smart Cities

The following are some of the application domains that need to be implemented in smart cities:

- *Transportation*: This section discusses how ICT can manage transportation, including intelligent transportation goods and mobility.
- *Urban infrastructure* denotes using data and communication technology to manage physical and building assets (ICT).

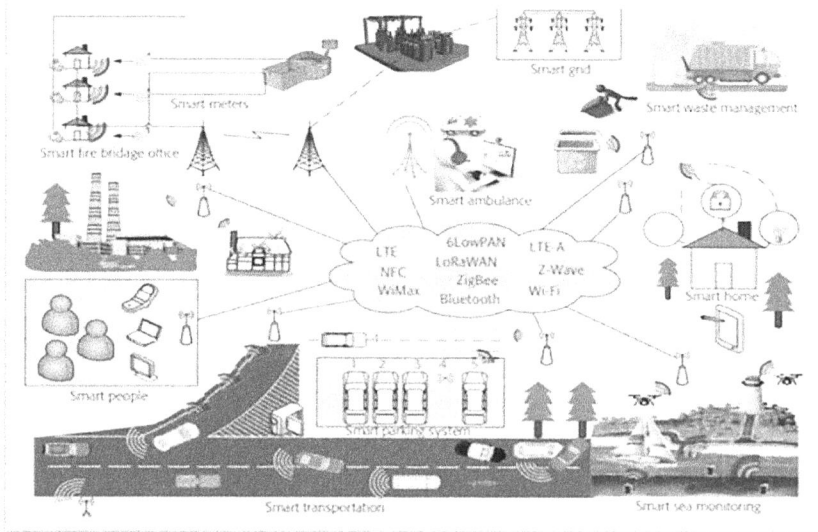

FIGURE 10.1 Allowing communication technologies for smart cities.

- *Living*: This term mentions the quality of an individual's life, as well as their education, health, and safety in a city.
- *Resource utilization and management*: The monitoring and management of natural resources, energy and water are all tied to the consumption and management of resources.
- *Government*: This refers to the functions of the government—for example, openness and the effectiveness of the city's administration.
- *Economy*: It includes aspects such as the city's domestic product, advanced essence, service, and e-commerce.

10.1.2 COMMUNICATION TECHNOLOGIES IN THE SMART CITY

High-speed data transmission, better network structure, small-signal loss, relatively low-priced, robust, authentic high quality and flexible encryption processes will, among others, be made possible by emerging and future technologies. Small cell technology has been proven to help smart city applications satisfy their communication and networking needs in terms of interoperability, robustness, and power consumption (Figure 10.1). For a better quality of experience, low power and multimodal access are required.

10.2 CRITICAL ASPECTS OF SMART CITIES

A smart city model has been created to measure a city in terms of economy, people, mobility, governance, environment and so on. Creating models can give people a better understanding of the city's current state, but it can also highlight the parts that need

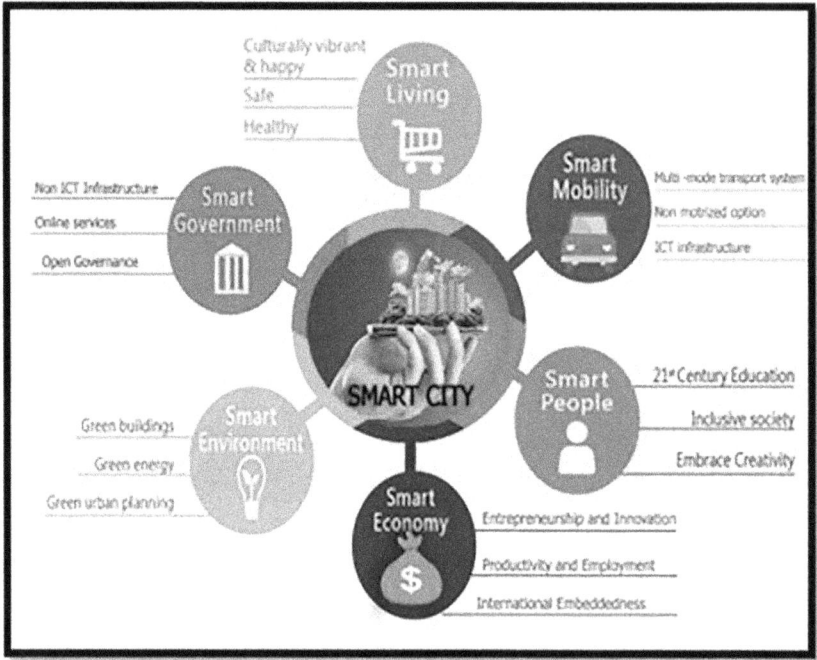

FIGURE 10.2   Overview of the various mechanisms that make up the six smart city features.

Source: Conference: 2019 Canadian Society for Civil Engineers at Laval (Greater Montreal), Canada.

improvement as a smart city [2]. Because of the specific characteristics of smart cities outlined earlier, designing, building, and maintaining intelligent systems has become extremely difficult. The specific requirements for developing an intelligent system are mostly determined by the nature of the applications and hardware and network infrastructure constraints. When designing intelligent systems, the designer will almost certainly have to deal with several specific and obvious challenges, such as public safety, healthcare, transportation and energy. A *smart city* is a forward-thinking metropolitan environment that serves businesses, institutions, and, most importantly, residents [3]. Even though smart cities encounter various trials during their growth and socioeconomic and political concerns, the most significant challenge is technical issues. Security and privacy are essential because of these technical difficulties and other problems, such as organization interoperability and cost-effective technology (Figure 10.2).

10.3   ARCHITECTURAL METHODS FOR SMART CITY DEVELOPMENT

This segment provides a short-term explanation of some proposed architectures for innovative city development over time. It also addresses the architecture layers, event-driven architecture (EDA), IoT, service-oriented architecture (SOA), and hybrid architecture.

10.3.1 ARCHITECTURAL LAYERS

The architectural layer employs a structure to construct applications and services in the smart city. Every single layer is materially and logically separated from the others. Architectural layers have this property, leading to researchers' universal recognition. There are some examples of work that has used the architectural layer concept for smart cities. One of the first in the industry is the author's online comparison of smart cities in the United States, Amsterdam, Helsinki and Kyoto. "Information layer" is the initial layer by this definition [4]. This layer combines real-time sensor data and Internet files, which are then merged using geographic information systems (GIS). The interface class is the second one. This division is responsible for building a virtual city setting using 3D space and 2D maps. The final layer is the interaction layer. In this scenario, an agent system is employed to communicate.

Carretero recently created a self-adaptive system for smart cities utilizing the ADAPCITY architecture. It gives various devices the ability to respond quickly in various situations. This technology is also capable of immediate recovery and the generation of new operations and upgrades. There are four layers to the architecture in this case. The state and conduct of gadgets and items are addressed in the principal layer, the actual layer. The subsequent layer, the passage layer, is accountable for handling, putting away, and sending information from the actual layer. The third level, the business layer, utilizes factual procedures, information mining, and expectation to deal with the information handled from the network layer. The last layer is the control layer, which incorporates the administrations provided while considering client inclinations and enhancement measurements.

10.3.2 SERVICE-ORIENTED ARCHITECTURE

The main goal of an SOA is the communication, collection and interaction between services. Among different services, communication in a computer system occurs by exchanging data between them. Each collaboration is considered unrestricted because the services are dissimilar, weakly coupled, and independent.

10.3.3 INTERNET OF THINGS

A few mixed gadgets are associated with the Internet and order themselves utilizing IP locations and conventions in the Internet of Things (IoT). The gadgets are encircled by sensors and actuators associated with the Internet remotely. Building intelligent systems for smart cities IoT likewise gives network and correspondence among sensors, permitting clients to get to various utilizations. The IoT-based design has been fundamental in the advancement of intelligent urban areas (Figure 10.3).

A framework for a smart city named smart city basic foundation was created by Attwood et al. [5]. The rationale of this framework is to shield basic foundations from disappointment. It likewise helps the framework improve and keep working if a disappointment is unavoidable. Sensor actuator networks are extremely significant for working this multitude of capacities easily. These actuator networks interface to IoT for get-together information required by the smart city. In [6], Asimakopoulou and Bessis discussed about the possibilities of crowd sourcing in the contexts of

IoT Components

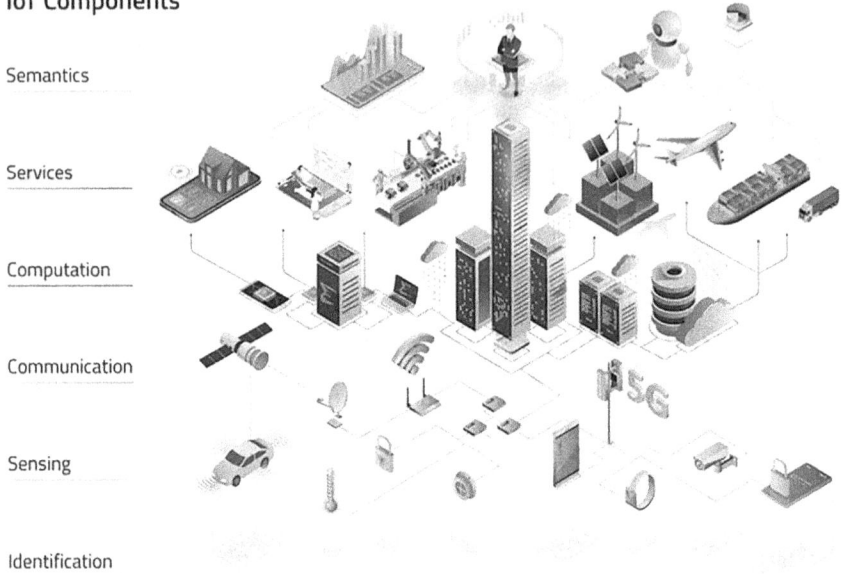

Semantics

Services

Computation

Communication

Sensing

Identification

FIGURE 10.3   Internet of Things (IoT) architecture components.

smart buildings and cities in order to support a more effective and efficient integrated disaster management approach. Residents with APIs on their cell phones utilize publicly supporting innovation to find crisis events and dangers. Network registering to coordinate heterogeneous assets, distributed computing to work with admittance to these assets, and unavoidable processing to gather and deal with information from gadgets are among the extra advancements supported by the creators. Wang et al. [7] exhibited the utilization of NASA World Wind, which is launched by National Aeronautics and Space Administration (NASA), an open-source geographic software, to reproduce a city in another work. The program is an open-source stage that empowers representation, reproduction, and connection in every aspect of a smart city's living quarters. Information gathering and visual showcase are the two critical parts of this innovation. The data is assembled through IoT, network investigation, and web map administrations. Samaras et al. [8] made the SEN2SOC smart city stage in Spain's Smart Santander City. The objective was to work on the way of life of individuals and guests in a smart city by expanding the cooperation among sensor and interpersonal organizations utilizing a characteristic language age (NLG) innovation. The SEN2SOC stage is based on a part-based plan containing portable and online applications, sensor and social information observing measurable examination, and a connection point. Horng proposes a smart parking system in [9]. The suggested method makes it quick and straightforward for individuals to find parking places, which helps to reduce traffic congestion and pollution. WSNs are used in the proposed smart parking system to detect the presence of automobiles approaching a parking space. An internal suggestion process of the specific location notifies the parking congestion cloud center in a smart parking system (PCCC). Aside from the architectures listed above, certain projects are carried out by integrating the features and technology of the various architectures.

10.4 REAL-LIFE APPLICATIONS

In smart urban communities, the interest in systems administration and correspondence innovation extends to give a wide scope of availability administrations. Smart urban communities depend entirely on network availability, which requires fast, high trustworthiness, and openness, yet the qualities that the present organizations request. This part gives a fast outline of some best-in-class works carried out in smart urban areas across an assortment of utilizations.

10.4.1 Food Management

This section describes researchers' research to solve real-world food management issues in smart cities. The food store network in present-day urban communities has extended quickly due to rising human requests. The IoT can be utilized to screen and deal with the food store network [10]. Some analysts investigated a situation by following the starting points of the food production network, which is a significant IoT application for smart urban areas. The specialists proposed a savvy sensor information assortment technique to screen, investigate, and manage the food area in urban communities. More specifically, the proposed data collecting technique is utilized to track back diseased food in the marketplaces and find the source of contamination.

10.4.2 Energy Management

Regarding the design of smart cities, energy management is a critical concern. Considering the current circumstance, the objective of brilliant city applications ought to be to advance the utilization of environmentally friendly power in their tasks. Additionally, accentuation should be paid to energy preservation, which requires decreased energy use. In [11], the creators proposed the plan of a sustainable energy microsystem (SEM), which intends to coordinate dissimilar freely working subsystems, like scattered age from inexhaustible and consolidated hotness and power units, re-energizing module half breed and electric vehicles utilized for surface portability and so on. In this work, the creators attempt to move past the idea of brilliant matrices, looking for arrangements that would consider a more incorporated guideline of energy streams between various subsystems, which will be a basic part of a smart city.

The SEM is envisioned as a flexible energy hub capable of supplying and storing a wide range of energy carriers. As a result, this paper presents new methods for optimizing SEM design and e-governance. The authors have also attempted to build innovative technology-based strategies for decreasing infrastructure vulnerabilities. They also looked at managing infrastructure interdependencies and the environment's impact on infrastructure. Finally, the study looks into organizational concerns and human factors influencing SEM control and management. The authors provide answers in a second study [12] through the DC4Cities program, which attempts to maximize the use of local renewable energy sources while running data centers in smart cities. Data centers, in general, are essential components of smart cities since they act as producers of IT services and energy consumers.

10.4.3 TRANSPORT MANAGEMENT

Intelligent transportation systems are also essential to a smart city [13]. One of the applications of smart cities, network idleness, might cause major problems in smart mobility. The purpose of traffic apps is to prevent fatal and property-damaging road accidents [14–15]. The sensor devices alert drivers to potential hazards. These devices interact with one another to alert drivers to situations beyond their field of vision. As a result, it aids in the prevention of serious accidents. These sensor devices work in two ways: one uses a specific approach, while the other uses a combination of methods. Whenever vital information is discovered, it is transported to a designated spot where passing vehicles will subsequently retrieve it. The second method is particularly useful for recognizing non-ephemeral events [16]. According to researchers, static sensor nodes should be placed at the beginning of every route.

In other research [17–18], the authors collaborate between two sensor nodes to identify speed limit violations in traffic law enforcement applications. Cameras are triggered when a speed violation is detected, and the images are sent to the traffic management center (TMC) for processing and storage. Before imposing any penalties, the drivers might be warned using variable message signs (VMSs) [18]. Illegal parking is detected by installing sensor nodes photographing the car's number plate that caused the problem. The author's employed strategies for post-accident investigation findings in another paper [19]. After an accident, a post-accident investigation is required to determine who is responsible. An application mentioned by the authors is traffic control at intersections using traffic lights, which allows traffic to be scheduled. Sensor nodes, usually one per lane, are mounted on traffic signals to determine the quantity of traffic arriving at each segment's intersection. After the traffic lights are installed, sensor nodes can be placed to determine the line length at each traffic light. All these methods require a limited number of sensor nodes, which reduces cost. Smart mobility is now getting more popular due to its portability feature [20].

10.5   MODERN COMMUNICATION TECHNOLOGIES

The primary motivation for changing a city into a smart city is to improve citizens' lives in many circumstances. Only when users have perfect connectivity is it likely to happen. Some of the smart city's precision-based promises call for high-quality communication technologies. Communication technologies allow diverse smart gadgets to communicate with one another. A wireless connection allows a mobile user to connect to a local network.

10.5.1 Mobility Management

E-health, intelligent transportation systems and logistics, to name a few, are available to mobile users in smart cities. These applications rely on services, from low data rate multimedia requests to real-time (high-speed) software applications. Spread across multiple access networks. Accordingly, the development of smart mobility approaches that leverage different wireless access technologies to enable global roaming is one of the top research priorities for mobile systems in the region. In addition, upcoming wireless technologies in smart cities will require integration and compatibility with current mobility management systems across different access networks.

10.5.2 High-Energy Consumption

Communication and networking from an energy perspective have received much attention. While new communication technologies such as WiMAX and LTEA have provided consumers with speedy upload and download speeds, their very high-power consumption rates can limit their usage. Even if future gadgets have superior battery life standards, present communication technologies' energy consumption rate will still be regarded as higher. Modern communication technologies consume a lot of energy for various reasons, including enhancing the radio network to achieve good signal quality.

10.6 ISSUES AND CHALLENGES FOR BUILDING INTELLIGENT SYSTEMS

This segment delves into smart cities' various concerns and challenges when planning, constructing and implementing intelligent systems.

10.6.1 Issues

There are numerous issues when discussing constructing intelligent systems for smart cities. These problems are technical as well as of social nature and include the following:

- *Transport*: Removing overcrowding, integrating all forms of transportation, and facilitating the issue of new avenues are all zones that must be considered while planning the transportation element in smart cities.
- *Public safety*: Real-time data analysis will be carried out to reduce misconduct and respond quickly to dangers to the public.
- *Healthcare*: More improved connectivity and innovative analytics for analyzing a significant quantity of data collected are required for better healthcare.
- *Energy issue*: Areas should be investigated where energy from traditional systems can be acquired and used to operate several requests in smart cities. Society should also be alert to their energy consumption and design strategies to reduce it.

10.6.2 CHALLENGES

Like any other field, smart cities encounter numerous significant issues that must be addressed while they are being developed.

- *Application cost*: The emergence of smart cities necessitates high-cost technologies. Many cities have spent a lot of money trying to install technology with the eventual goal of making the city "smart."
- *Mobility*: In smart cities, it's critical to provide mobile consumers with uninterrupted service when switching between different access networks. In smart cities, it's critical to ensure that applications work smoothly.
- *Scalability*: The limitations in storing, bandwidth, and computational powers that are a disadvantage to service providers when dealing with many operators should not prevent the correct operation of the many amenities valid to smart cities.
- *Technology integration*: Smart cities quickly gain traction as a feasible solution for enhancing prospects. While ICTs are used to establish smart cities that use calculations and data, they are also used to develop such cities in terms of resources and structure. As a result, demonstrating how the technologies used in the advancements mentioned earlier are interoperable is a critical task if cities are truly called smart.
- *Upgrading*: Any technology, including smart cities, requires this characteristic to advance. Because the city relies on communication and technology, upgrading a smart city will be costly, posing a significant problem.

10.6.3 STATE-OF-THE-ART SOLUTIONS

This section summarizes the several crucial concerns that have been solved in the current development of intelligent systems for smart cities [21]. Suryadevara created an affordable, adaptable, resilient, and smart data control system to detect the health of elderly people living alone in the smart home. WSN and smart home monitoring software systems are two components that make up the model of the system. These modules are in charge of gathering sensor data and analyzing data to detect changes in older behavioral patterns. Healthcare facility providers help the elderly based on their behavioral patterns. The writers tested the models in many elderly homes, with compelling results [22]. Similarly, Dasios et al. developed a WSN-based intelligent senior care monitoring system. Initially, the system tracks their everyday behaviors, such as walking, sitting, and sleeping, by monitoring and recording many environmental factors. The system sends automated alarms to authorized people whenever any major change from an individual's usual activity pattern is detected. An intelligent vehicle speed control system is provided [23]. To manage vehicle speed, RFID is used in the suggested system to communicate between the car and traffic signs [24]. Magpantay et al. developed and installed a WSN to track electric energy use in smart buildings in a separate study. In this case, the granular radio energy (GRE) sensor node is employed to generate WSN. To reduce power consumption, the authors obtained positive results using WSN-based GRE sensing nodes in smart buildings [25]. Kim and his colleagues developed EnerISS, the equivalent of a smart energy monitoring system. EnerISS collects energy usage statistics from buildings using WSN. Collected energy usage data is kept in the EnerISS database. Ultimately, EnerISS makes timely decisions to effectively manage the energy supply grounded on the rules created by the contact between consumers and suppliers. Sensor nodes are employed in the proposed system to monitor three specific parameters: vibration, pressure, and sound. The system effectively senses a pipeline leak by monitoring these characteristics.

10.6.4 FUTURE DIRECTIONS

There have been several studies on developing intelligent systems for smart cities. However, some major exposed challenges have yet to be answered or thoroughly studied. This section overviews the study fields for developing intelligent systems for smart cities. The following are some of the areas where further improvements can be made:

- *Cloud platform development*: The use of cloud computing platforms in smart cities is one topic that needs to be investigated in the future, as cloud services can make life easier for smart cities. This can be done by ensuring public safety and security through traffic policy or regulation, coordinated public works, city maintenance and other methods. As a result, experts are interested in the issues of using cloud services to manage vast amounts of data in smart cities.
- *Incorporating IoT into a larger picture*: Even though the IoT has become a vital part of smart city operations, more work is still needed to utilize IoT in these cities effectively. The IoT is crucial for retrieving relevant info from actual data. The themes covered by IoT services are diverse. It is possible to conduct more research into the role of IoT in smart cities.
- *Safeguarding interoperability*: Several city systems that practice cutting-edge security, such as lighting, transit and others, allow free data mobility. As a result, intelligent communication between such technologies is possible. It is thus required to ensure compatibility amongst such systems, which could be pursued as a future research project.
- *Data management*: Data is crucial in creating a smart city. Smart cities will create a great amount of data; therefore, managing and comprehending it will be a significant problem. Furthermore, a large amount of storage is essential to store such a large amount of data. There needs to be more synergy between the application and implementation perspectives to address these issues.

10.7 CONCLUSION

Smart city services are expected to become crucial in the not-too-distant future. Nonetheless, much more work must be done, and existing frameworks must be refined. Many intelligent devices in smart cities have resulted in various connectivity issues, which may obstruct the implementation of existing communication technologies. Considering all of these aspects, we created this poll in which the general notion of smart cities and concerns and obstacles are examined.

Furthermore, how intelligent systems, such as transportation, play an important role is highlighted by addressing research projects. Smart city development is hampered by several factors, including social, political and, most importantly, technical challenges. The technological difficulties concern system compatibility and cost-effective technologies, focusing on security and privacy concerns. Because the networks that are a significant part of the operation of smart cities are vulnerable to hostile attackers, security is a top priority. As a result, providing unrestricted connections in smart cities will require much effort in the future.

REFERENCES

[1]. Caragliu, A., Del Bo, C., & Nijkamp, P. (2013). Smart cities in Europe. In *Smart cities* (pp. 185–207). Routledge.

[2]. Giffinger, R., & Gudrun, H. (2010). Smart cities ranking: An effective instrument for the positioning of the cities? *ACE: Architecture, City and Environment, 4*(12), 7–26.

[3]. Khatoun, R., & Zeadally, S. (2016). Smart cities: Concepts, architectures, research opportunities. *Communications of the ACM, 59*(8), 46–57.

[4]. Kubina, M., Šulyová, D., & Vodák, J. (2021). Comparison of smart city standards, implementation and cluster models of cities in North America and Europe. *Sustainability, 13*(6), 3120.

[5]. Attwood, A., Merabti, M., Fergus, P., & Abuelmaatti, O. (2011). SCCIR: Smart cities critical infrastructure response framework. In *2011 Developments in E-Systems Engineering* (pp. 460–464). IEEE.

[6]. Asimakopoulou, E., & Bessis, N. (2011). Buildings and crowds: Forming smart cities for more effective disaster management. In *2011 Fifth International Conference on Innovative Mobile and Internet Services in Ubiquitous Computing* (pp. 229–234). IEEE.

[7]. Wang, R., Jin, L., Xiao, R., Guo, S., & Li, S. (2012). 3D reconstruction and interaction for smart city based on World Wind. In *2012 International Conference on Audio, Language and Image Processing* (pp. 953–956). IEEE.

[8]. Samaras, C., Vakali, A., Giatsoglou, M., Chatzakou, D., & Angelis, L. (2013). Requirements and architecture design principles for a smart city experiment with sensor and social networks integration. In *Proceedings of the 17th Panhellenic Conference on Informatics* (pp. 327–334). Association for Computing Machinery.

[9]. Horng, G.J. (2015). The adaptive recommendation mechanism for distributed parking service in smart city. *Wireless Personal Communications, 80*(1), 395–413.

[10]. Zhang, Q., Huang, T., Zhu, Y., & Qiu, M. (2013). A case study of sensor data collection and analysis in smart city: provenance in smart food supply chain. *International Journal of Distributed Sensor Networks, 9*(11), 382132.

[11]. Brenna, M., Falvo, M. C., Foiadelli, F., Martirano, L., Massaro, F., Poli, D.A.V.I.D.E., & Vaccaro, A. (2012). Challenges in energy systems for the smart-cities of the future. In *2012 IEEE International Energy Conference and Exhibition (ENERGYCON)* (pp. 755–762). IEEE.

[12]. Klingert, S., Niedermeier, F., Dupont, C., Giuliani, G., Schulze, T., & de Meer, H. (2015). Renewable energy-aware data centre operations for smart cities the DC4Cities approach. In *2015 International Conference on Smart Cities and Green ICT Systems (SMARTGREENS)* (pp. 1–9). IEEE.

[13]. Losilla, F., Garcia-Sanchez, A.J., Garcia-Sanchez, F., Garcia-Haro, J., & Haas, Z.J. (2011). A comprehensive approach to WSN-based ITS applications: A survey. *Sensors, 11*(11), 10220–10265.

[14]. Birk, W., Osipov, E., & Eliasson, J. (2009). iRoad-cooperative road infrastructure systems for driver support. In *World Congress and Exhibition on Intelligent Transport Systems and Services: 21/09/2009–25/09/2009*. Curran Associates, Inc.

[15]. Qin, H., Li, Z., Wang, Y., Lu, X., Zhang, W., & Wang, G. (2010, March). An integrated network of roadside sensors and vehicles for driving safety: Concept, design and experiments. In *2010 IEEE International Conference on Pervasive Computing and Communications (PerCom)* (pp. 79–87). IEEE.

[16]. Kong, F., & Tan, J. (2008, June). A collaboration-based hybrid vehicular sensor network architecture. In *2008 International Conference on Information and Automation* (pp. 584–589). IEEE.

[17]. Yoo, S.E., Chong, P.K., & Kim, D. (2009). S3: School zone safety system based on wireless sensor network. *Sensors, 9*(8), 5968–5988.

[18]. Bohli, J.M., Hessler, A., Ugus, O., & Westhoff, D. (2008). A secure and resilient WSN roadside architecture for intelligent transport systems. In *Proceedings of the First ACM Conference on Wireless Network Security* (pp. 161–171). Association for Computing Machinery

[19]. Chang, Y.S., Juang, T.Y., & Su, C.Y. (2008). Wireless sensor network assisted dynamic path planning for transportation systems. In *International Conference on Autonomic and Trusted Computing* (pp. 615–628). Springer.

[20]. Wanganoo, L., Shukla, V., & Mohan, V. (2022). Intelligent micro-mobility e-scooter: Revolutionizing urban transport. *Trust-Based Communication Systems for Internet of Things Applications*, 267–290.

[21]. Suryadevara, N.K., Mukhopadhyay, S.C., Wang, R., & Rayudu, R.K. (2013). Forecasting the behavior of an elderly using wireless sensors data in a smart home. *Engineering Applications of Artificial Intelligence, 26*(10), 2641–2652.

[22]. Dasios, A., Gavalas, D., Pantziou, G., & Konstantopoulos, C. (2015, July). Wireless sensor network deployment for remote elderly care monitoring. In *Proceedings of the 8th ACM International Conference on Pervasive Technologies Related to Assistive Environments* (pp. 1–4). Association for Computing Machinery.

[23]. Pérez, J., Seco, F., Milanés, V., Jiménez, A., Díaz, J. C., & De Pedro, T. (2010). An RFID-based intelligent vehicle speed controller using active traffic signals. *Sensors, 10*(6), 5872–5887.

[24]. Magpantay, P., Paprotny, I., Send, R., Xu, Q., Sherman, C., Alarcon, L., . . . & Wright, P. (2014). Energy monitoring in smart buildings using wireless sensor networks. In *Proceedings 3rd International Conference on Smart Systems Devices and Technologies* (pp. 78–81). IARIA XPS Press, ISSN: 2308-3727, ISBN: 978-1-61208-363-6.

[25]. Kim, S. A., Shin, D., Choe, Y., Seibert, T., & Walz, S. P. (2012). Integrated energy monitoring and visualization system for Smart Green City development: Designing a spatial information integrated energy monitoring model in the context of massive data management on a web based platform. *Automation in Construction, 22*, 51–59.

Part IV

---

*Computational Intelligence
Supporting Urban Infrastructure*

11 Addressing Smart Fire Detection Management through the Internet of Things and Arduino

Sameera Ibrahim, Vinod Kumar Shukla,
Shaurya Gupta and Purushottam Sharma

CONTENTS

11.1 INTRODUCTION

The IoT presents the world with many advantages and can deliver the services to create large numbers of applications and help to introduce them to the public forum. The IoT technologies offer a mild and less costly means of detecting and manipulating devices to incorporate IoT in domestic and industrial norms, where individuals tend to have power over items using smartphones that will make their everyday existence simpler [1]. IoT is the programming and networking of items on the Internet to make devices accessible and identifiable. The enhancement and transformation in fire monitoring using information and networking technology has strategic relevance for locations where fire incidents are inevitable. Compared to conventional fire detection systems, the automated fire detection and warning system (FDAS) platform is a promising, sustainable leading technology to identify fires effectively. The task of fire detection and warning systems is to recognize the development of an emergency as soon as possible and also to alert the residents of the building. Fire detectors and alarm systems can support many core functions.

DOI: 10.1201/9781003218715-15

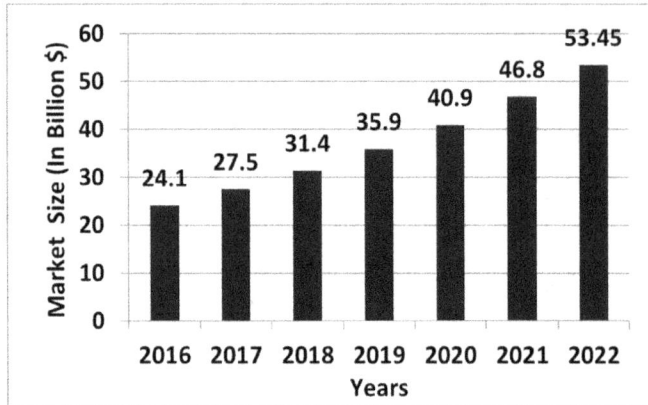

FIGURE 11.1   Global smart home market from 2016 to 2022 (in billion USD).

11.1.1 GROWTH OF IoT MARKET

As depicted by Figure 11.1 [2], the global smart IoT market has seen a significant increase in the years 2016–2020 and similar growth is projected for the year 2021–2022. IoT-based smart homes allow users to control their home lighting systems, air conditioners and so on without switching them on and off using physical switches. Hence this will enable remote controlling of these devices as these devices are Internet connected. Similarly, IoT-based systems are used in smart homes to help in intelligent healthcare monitoring [3–5].

IoT is playing a vital role in various domains such as tourism [6–7], food, hospitality and service industry [8–9], supply chain [10], office management [11] and the transport sector with personalization [12–14], to name a few.

There is a need for devices for autonomous flame detection. These devices perform the functions of quick detection, providing a warning notification, and fire-side termination initiation. The mechanisms, temperature sensors fitted with smoke, can detect unfavorable incidental conditions as they unfold. With the assistance of a processing unit, it will automatically alert you to carry out cautious steps. Early detection and a faster alarm will result in fewer losses [15].

Sensors are an important part of an environment that is smart and self-decisive. Sensors in a community of topology and architecture form wireless sensor networks suitable for a system commonly used in health tracking, emergency control, fire detection, intrusion detection and so on [16].

Concerning the advantages of wireless sensors, the network is still considered a cheap commodity in terms of equipment and installation over conventional fire outbreak protection systems. As IoT technology is becoming increasingly popular in the commercial market, it is becoming more attractive for its related systems and components, including the wireless sensor network used in protection and, in our case, fire safety [17].

The wireless sensor network is considered a practical strategy for fire outbreak protection systems that have recently attracted considerable attention and have been well developed. It is considered a sudden occurrence that needs a predictable safety solution to handle this kind of danger and is likely to occur anytime and anywhere [1].

Fire sensors that are considered part of the wireless sensor network play an important role in monitoring and detecting any irregular rise in temperature and humidity rates. In this chapter, a sensor is used to track the temperature and humidity rates in a particular area; this sensor is connected to an Arduino node MCU chip; this chip is connected to a Wi-Fi network, and the node MCU will constantly submit information from the sensor to an Internet-connected database, the node MCU would then receive these changes from the sensor as data and send it to a database via the Internet that will be stored there as values. These values will be sent by the database to our mobile phones as an SMS message to inform us about the incident; it can also be used to send a warning to fire stations and hospitals in the case of factory fires, forest fires and fires at homes [18].

This model employs Arduino fundamentals and a flame sensor to simplify the way we monitor any damage. This is done by interfacing the flame sensor with the Arduino Uno microcontroller-based device. It burns surrounding objects once the flame is ignited. The flame sensor module detects the surrounding temperature and clicks on the buzzer attached to the Arduino Uno output via the relay. We have attempted to make it more intelligent and regulated by Arduino programming with the support of IoT technologies.

11.2   LITERATURE REVIEW

The significance of an IoT-based smart home system is to help detect smoke and alert authorized personnel of the cause by SMS. These sensors can be configured in a manner that these sensors only work when we are not physically present in the home. These sensors are capable of detecting smoke when there is a sign of visible carbon suspension that is released from a matter that is being burned. The importance of such smoke detectors is very important as indoor deaths happen when there is a fire breakout, and victims inside apartments or flats with fire breakout inhale the smoke as it is toxic and lead to carbon monoxide poisoning as the most dangerous component of smoke is carbon monoxide [19]. False smoke alarms are common and may cause an incident to cost around $30,000 to $50,000 [20]. These smoke detection systems also use send the signals to the main control system, which not only shows the smoke alerts but also other regular data like the temperature of various rooms, and this will help to send these signals to the central main controller, which makes the user with access to the controller monitor and control certain activities of the room [21].

Fire detection has been a big issue, and the introduction of IoT and WSN technologies has helped the detection of fire much easier. IoT sensors ensure that data collected from fire sensors send alarms or alerts to a remote server that may be alerted by the authorities with the mobile application installed. These data might be in the form of temperature collected by the temperature and fire sensor [22]. Such a smart detection and alerting system are only possible with IoT as it has enabled all these devices or things to be connected to the Internet to monitor devices connected to the Internet and also to monitor the data generated by them and also helps a user to control them remotely making all these features available for a very lower cost [23]. WSN's makes it possible for these data collected by IoT sensors to be transmitted wirelessly to the Internet. ZigBee is used popularly for wireless communication among sensor nodes [24].

11.3   FRAMEWORK OF PROPOSED MODEL

In commercial buildings and warehouses, fire alarm systems are quite popular; such devices typically contain a group of sensors that continuously check for any flame or fire in the house. If any of these are detected, it will cause an alarm. Using an IR flame detector, these devices have an IR photodiode that is reactive to IR light, one of the easiest ways to detect flames. Since fire creates heat in the event of a fire and releases IR rays, this light is not noticeable to human eyes. However, the fire sensor will sense it and notify a microcontroller such as Arduino that a fire has been identified.

In this chapter, we have integrated the flame sensor with Arduino and explored the steps to create a fire alarm device using Arduino and flame sensor. The flame sensor module has a photodiode for light detection and an operational amplifier for sensitivity function. It is used to identify fire upon detection and displays a strong signal. The signal is read and warned by Arduino by switching the buzzer and LED on. An IR-based flame sensor is used. The core component of the hardware circuit is the microcontroller, which monitors and allows the whole circuit to operate. It is quick to write code and submit it to the board with the help of open-source Arduino software (IDE). When the sensor provides output, the microcontroller switches on the relay circuit that is further linked to the buzzer and LED light; therefore, when the sensor senses the flame, the buzzer is triggered, and the LED turns on. In addition to the above, blaze detection was carried out only to make it easy for the fire protection to find the susceptible area as quickly as possible [25].

Modern fire warning devices use automatic features to predict the risk of an incident that can result in a fire. They acquire a sign from a sensing feature of the fireplace (smoke, heat detector) and relay it to the fireplace warning panel mechanically. Radio frequency in wireless devices conveys the sensor signal to the control center. Heat detectors also have built-in thermistors to feel the temperature.

11.4   HARDWARE USED

11.4.1   Arduino

Arduino has a basic microcontroller board, and the related open-source physical computing system provides a programming environment for writing code for the board. It is an open-source framework used to develop projects for electronics. Arduino consists of a physical programmable circuit board called a microcontroller and a set of software running on the device that is used to write and upload code to the IDE [26].

11.4.2   Flame Sensor

The flame sensor is a device which could be used to sense the presence of the source of fire or some other source of strong light. To implement a flame sensor, an infrared radiation-sensitive sensor is used. The KY-026 flame sensor module for Arduino senses fire-emitting infrared radiation. To change the resistance, the module also has digital and analog outputs and a potentiometer. It is used widely in applications of fire detection [26].

11.4.3 BUZZER

A buzzer is an audio signaling system that is electronic, electromechanical, electro-magnetic or piezoelectric. An oscillatory electrical device or another audio signal source may drive a piezoelectric buzzer. It consists of various switches or sensors that are attached to a control unit and decide whether and which button has been pressed or a predetermined time has passed. Upon this, the control panel typically turns the light on and sounds an alarm in the form of a constant buzzing or beeping [26–27].

11.5 BENEFITS OF FIRE DETECTION SYSTEM

The warning advantage is the primary benefit of implementing fire prevention and alarm systems. Installing fire warnings and alarm systems anywhere in a house is possible. For successful fire protection, early notice is crucial since fires can occur anywhere and anywhere. This then helps to:

- Improve the evacuation time for building residents before a fire spreads out of control.
- Deploy medical rescue support to those in need.
- Allow staff from the fire service to quickly respond.

The development of the proposed model (Figure 11.2) introduces a fire prevention prototype and an alert system that provides residents with a notification alert whenever the value of the sensor exceeds its threshold value. This architecture contains a flame sensor, LED light, and buzzer and transmits constructed and programmed into an Arduino microcontroller to develop an enabling fire safety system.

In terms of its functionality and layout, the existing fire alarm system on the market today is too complicated. Since the mechanism is so complicated, it requires constant maintenance to be carried out to ensure that the system performs properly.

We have to know the basic device design of the IoT-enabled system and flame sensor to understand the function of the project. The flame sensor is used to position at certain distances, so it senses the flame (heat). A flame sensor detects a poor DC signal from the AC power sent to the ignitor that is corrected to DC by the flame rectification process by which the power circuitry sent through a flame is rectified. In this research paper, this sensor was used to sense the fire. The sensor sends the signal or input to the microcontroller. Both of these can detect environmental changes and respond immediately after emergencies.

This chapter uses an Arduino circuit to design and implement a fire detection system. The Arduino controller serves as the central system, using the flame sensor to detect temperature increases and heat.

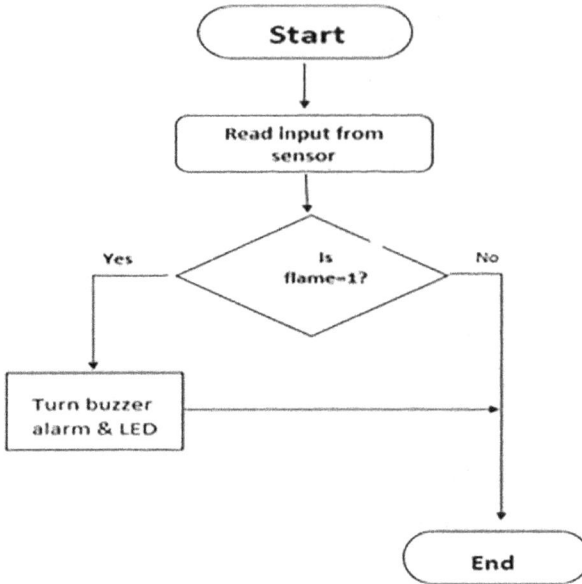

FIGURE 11.2 Proposed prototype for fire detection.

The buzzer acts as an alert system, and wireless communication among Android phones and Arduino is created using the Wi-Fi module. The infrared flame sensor will constantly check the increase in temperature and bright light sources. The results will be provided to the controller from the sensor. If the fire is detected or the temperature increases, the information is provided to the controller, and the buzzer is activated to notify the controller. The controller would also give a message for the detection of fire and smoke via the Arduino microcontroller to the Android phone virtual terminal.

After making the appropriate connections, the code is uploaded into the Arduino Uno. Then, a fire lighter can be placed next to the sensor to verify the flame sensor's responsiveness.

The output from the flame sensor is high under ordinary circumstances. When the sensor senses fire, its output becomes low. The Arduino senses this low signal on its input pin and triggers the buzzer.

11.6 APPLICATIONS OF IOT/FIRE DETECTION

- IoT-based smart building systems have sophisticated building management systems that timely alert fire breakout, which can help faster fire detection in buildings [28].
- IoT-based fire detection in the forest where IoT sensors detect various parameters like smoke, temperature and so on and these data generated are then transmitted wirelessly using popular wireless technologies like ZigBee, 4G, WLAN and so on [29].

- Unmanned aerial vehicles being used for fire detection and firefighting have helped both day and night to help fight the fire and to detect them at very reasonable costs [30].

11.7 CONCLUSION

This method introduces the development of a fire alarm system using the Uno Arduino. This device undoes the need for an individual to control the environment on an ongoing basis. With the assistance of sensors, tracking will be performed. In order to warn the appropriate entities, buzzer and notification warnings are used. This system is a low-cost, energy-efficient system focused on accurate and robust instruments.

Incorporating IoT into the fire protection system dramatically improves its reliability and efficacy. Fire signals such as temperature rise and the appearance of flames, fumes and smoke are easily monitored using sensors. Via alert sound, light warnings and SMS messages received by the modules built into this device, building tenants and firefighting officials are alerted in real time. Critical circumstances are solved and resolved easily in conventional systems that take a significant amount of time and resources [31]. This wireless fire detection system architecture and installation is adaptive and versatile. This approach for wireless detection is more cost-effective than the fire detection devices present on the market. This particular flame detection system has a high accuracy rate, identifies temperature degree differences, and provides strong security. The adaptability of this system is one of the basic points of concern.

In this research paper, an effort is made to develop a fire alarm device for the effective use of electricity using a flame sensor and a microcontroller. It would help to minimize energy consumption, save lives, reduce the number of injuries and reduce electrical appliance waste. The findings obtained from the measurement showed that, under all conditions, the device works well. The key aim of this project was to build a high-temperature circuit that activates an alarm to help individual's sense fire and extinguish the fire as soon as possible.

REFERENCES

[1]. Sharma, S., Chand, K., Sharma, D. and Guha, P., Development of an early detection system for fire using Wireless Sensor Networks and Arduino. In *2018 International Conference on Sustainable Energy, Electronics, and Computing Systems (SEEMS)*, 2018, October, pp. 1–5. IEEE.

[2]. "Forecast market size of the global smart home market from 2016 to 2022 (in billion U.S. dollars),"[online] www.statista.com/statistics/682204/global-smart-home-market-size/#:~:text=The%20total%20installed%20base%20of,be%20connected%2Fsmart%20home%20devices. [Accessed on: 07–03–2021]

[3]. Ibrahim, S., Shukla, V. K. and Bathla, R. Security enhancement in smart home management through multimodal biometric and passcode. *2020 International Conference on Intelligent Engineering and Management (ICIEM)*, 2020, pp. 420–424, doi:10.1109/ICIEM48762.2020.9160331.

[4]. Shukla, V. K. and Verma, A. Enhancing user navigation experience, object identification and surface depth detection for "Low Vision" with proposed electronic cane. *2019 Advances in Science and Engineering Technology International Conferences (ASET)*, 2019, pp. 1–5. Dubai. doi:10.1109/ICASET.2019.8714213.

[5]. Shukla, V. K. and Verma, A. Model for user customization in wearable virtual reality devices with IoT for "Low Vision". *2019 Amity International Conference on Artificial Intelligence (AICAI)*, 2019, pp. 806–810. Dubai. doi:10.1109/AICAI.2019.8701386.

[6]. Shukla, V. K. and Verma, A. Analyzing the influence of IoT in tourism industry (February 23, 2019). *Proceedings of International Conference on Sustainable Computing in Science, Technology and Management (SUSCOM)*, Amity University Rajasthan, Jaipur—India, February 26–28, 2019, Available at SSRN: https://ssrn.com/abstract=3358168 or https://doi.org/10.2139/ssrn.3358168

[7]. Verma, A., Shukla, V. K. and Sharma, R., Convergence of IOT in tourism industry: A pragmatic analysis. In *Journal of Physics: Conference Series*, Vol. 1714, No. 1, 2021, p. 012037. IOP Publishing.

[8]. Grobbelaar, W., Verma, A. and Shukla, V. K., Analyzing human robotic interaction in the food industry. In *Journal of Physics: Conference Series*, Vol. 1714, No. 1, 2021, p. 012032. IOP Publishing.

[9]. Albuquerque, D. D., Shukla, V. K., Verma, A., Tyagi, S. K. and Sharma, P. Enhancing sustainable customer dining experience through QR code and geo-fencing. *2020 International Conference on Computation, Automation and Knowledge Management (ICCAKM)*, Dubai, 2020, pp. 190–196. doi:10.1109/ICCAKM46823.2020.9051470.

[10]. Wanganoo, L. and Shukla, V. K. Real-time data monitoring in cold supply chain through NB-IoT. *2020 11th International Conference on Computing, Communication and Networking Technologies (ICCCNT)*, 2020, pp. 1–6. doi:10.1109/ICCCNT49239.2020.9225360.

[11]. Shukla, V. K. and N. Bhandari, Conceptual framework for enhancing payroll management and attendance monitoring system through RFID and biometric. *2019 Amity International Conference on Artificial Intelligence (AICAI)*, Dubai, 2019, pp. 188–192, doi:10.1109/AICAI.2019.8701316.

[12]. Madana, A. L. and Shukla, V. K. Conformity of accident detection using drones and vibration sensor. *2020 8th International Conference on Reliability, Infocom Technologies and Optimization (Trends and Future Directions) (ICRITO)*, Noida, 2020, pp. 192–197. doi:10.1109/ICRITO48877.2020.9197783.

[13]. Siraj, A. and Shukla, V. K. Framework for personalized car parking system using proximity sensor. *2020 8th International Conference on Reliability, Infocom Technologies and Optimization (Trends and Future Directions) (ICRITO)*, Noida, 2020, pp. 198–202. doi:10.1109/ICRITO48877.2020.9197853.

[14]. Wanganoo, L. and Shukla, V. K. Real-time data monitoring in cold supply chain through NB-IoT. *2020 11th International Conference on Computing, Communication and Networking Technologies (ICCCNT)*, Kharagpur, 2020, pp. 1–6, doi:10.1109/ICCCNT49239.2020.9225360.

[15]. Perilla, F. S., Villanueva Jr, G. R., Cacanindin, N. M. and Palaoag, T. D., 2018, February. Fire safety and alert system using Arduino sensors with IoT integration. *Proceedings of the 2018 7th International Conference on Software and Computer Applications*, pp. 199–203. IEEE.

[16]. Habib, M. R., Khan, N., Ahmed, K., Kiran, M. R., Asif, A. K. M., Bhuiyan, M. I. and Farrok, O. Quick fire sensing model and extinguishing by using an Arduino based fire protection device. In *2019 5th International Conference on Advances in Electrical Engineering (ICAEE)*, 2019, September, pp. 435–439. IEEE.

[17]. Kang, J., Basnet, S. and Farhad, S.M., 2019, July. Smart fire-alarm system for home. In *Conference on Complex, Intelligent, and Software Intensive Systems*, pp. 474–483. Springer.

[18]. Khalaf, O.I., Abdulsahib, G.M. and Zghair, N.A.K., 2019. IOT fire detection system using sensor with Arduino. AUS, 26, pp. 74–78.

[19]. Philip Kpae, F.O., Macmammah, M., Ezekiel, N. D., Woji, C. T., Abubakar, A., & Kpaa Friday O.N., Design and Implementation of a Smart Home (Smoke, Fire, Gas and Motion Detector). *International Research Journal of Engineering and Technology (IRJET)*, 5(8). www.irjet.net/archives/V5/i8/IRJET-V5I8194.pdf [Accessed on: 07–03–2021]

[20]. Saeed, F., Paul, A., Rehman, A., Hong, W.H. and Seo, H., 2018. IoT-based intelligent modeling of smart home environment for fire prevention and safety. *Journal of Sensor and Actuator Networks*, 7(1), p. 11.

[21]. Rajesh, S., Kumar Thakur, A., Gehlot, A. A. Internet of things based on home automation for intrusion detection, smoke detection, smart appliance and lighting control. *International Journal of Scientific & Technology Research*, 2019. 8(12) [online] www.ijstr.org/final-print/dec2019/Internet-Of-Things-Based-On-Home-Automation-For-Intrusion-Detection-Smoke-Detection-Smart-Appliance-And-Lighting-Control.pdf [Accessed on: 07–03–2021]

[22]. Listyorini, T. and Rahim, R., 2018. A prototype fire detection implemented using the Internet of Things and fuzzy logic. *World Trans. Eng. Technol. Educ*, 16(1), pp. 42–46.

[23]. Sharma, A., Singh, P.K. and Kumar, Y., 2020. An integrated fire detection system using IoT and image processing technique for smart cities. *Sustainable Cities and Society*, 61, p. 102332.

[24]. Ryu, C.S., 2015. IoT-based intelligent for fire emergency response systems. *International Journal of Smart Home*, 9(3), pp. 161–168.

[25]. Mahzan, N.N., Enzai, N.M., Zin, N.M. and Noh, K.S.S.K.M., 2018, June. Design of an Arduino-based home fire alarm system with GSM module. *Journal of Physics: Conference Series*, 1019(1), p. 012079.

[26]. Myint, H. and Tun, M.Z., 2020, April, *Arduino Based Fire Detection and Alarm System Using Smoke Sensor* (Doctoral dissertation, MERAL Portal). *International Journal of Advances in Scientific Research and Engineering (IJASRE)*, 6(4), DOI: 10.31695/IJASRE.2020.33792

[27]. Dauda, M.S. and Toro, U.S., 2020. Arduino Based Fire Detection and Control System. *International Journal of Engineering Applied Sciences and Technology*, 4(11), pp. 447–453. Usman Saleh Toro.

[28]. Minoli, D., Sohraby, K. and Occhiogrosso, B., 2017. IoT considerations, requirements, and architectures for smart buildings—Energy optimization and next-generation building management systems. *IEEE Internet of Things Journal*, 4(1), pp. 269–283.

[29]. Neumann, G.B., De Almeida, V.P. and Endler, M., 2018, June. Smart Forests: fire detection service. *2018 IEEE symposium on computers and communications (ISCC)* (pp. 01276–01279). IEEE.

[30]. Kalatzis, N., Avgeris, M., Dechouniotis, D., Papadakis-Vlachopapadopoulos, K., Roussaki, I. and Papavassiliou, S., Edge computing in IoT ecosystems for UAV-enabled early fire detection. In *2018 IEEE International Conference on Smart Computing (SMARTCOMP)*, 2018, June, pp. 106–114. IEEE.

[31]. Chung, B.C. and Na, W., 2016. A study on the smart fire detection system using the wireless communication. *Journal of Convergence Society for SMB*, 6(3), pp. 37–41.

12 Development of Internet of Things Products for Smart Cities

Sonali P. Banerjee, Prasoon Banerjee, Rahul Gupta, Sunetra Saha and Deepak Jain

CONTENTS

12.1 INTRODUCTION

12.1.1 SMART CITIES

Various technical and commercial facilities and services are available in smart cities. These services are now supplied through closely linked, specific-domain systems, which hinders scalability and flexibility. The proposal for web-service delivery overcomes boundaries and encourages the introduction of new offerings based on an independent domain, an online service delivery platform offering real-time services. Services in smart cities are offered through a single domain, such as transportation management, building management and so on. These facilities are utilized for specific applications in that domain that require all system components to be driven and technical concepts to be determined, ranging from middleware components to sensors and smart devices, as well as computing infrastructure. Domain service provider to solutions for creating applications and multiple suppliers in integrating subsystems, and the service delivery process is strictly managed. The vertical system leads to a paradigm with closely connected hardware, networks, middleware and diverse application logic that can be closed.

DOI: 10.1201/9781003218715-16

In such systems, scalability and extensibility are constrained, and stakeholders impede the development of new services between closed connections. Web-scale services are provided for smart cities that have chosen a new methodology intending to offer smart services through open access in a scalable manner in collaboration with various stakeholders via cloud computing and the use of the Internet of Things (IoT). These services are offered through online software and use various applications to split the closed models of vertical deliveries.

12.1.2 INTERNET OF THINGS

The networking of humans, and objects with the support of the Internet, may exchange data across a network without physical contact among system to human and human to the system. It provides various levels of availability for devices and frameworks that operate with device-to-device communication (M2M) and are distributed across applications, domains and standards. Through remote Internet monitoring, there are various applications for IoT devices, such as heart monitoring implants, vehicles with sensors, smart thermostat systems, and clothes washers and dryers.

A smart city operates with the support of the IoT, which will increase the use of open assets; citizens will receive the enhanced quality of services from service providers at an affordable cost. This creates a unique environment for humans to enjoy as they wish, in the absence of explicit rules. The IoT connects smart devices using upgraded technologies, like radio frequency, sensors, actuators and other computer-to-computer connecting devices. Various combinations are used for business purposes. The IoT interface is built on three structural squares (some act like interaction). Communication is the main point of the Internet of information and data, rather than the other way around.

12.1.3 IoT PRODUCTS FOR SMART CITIES

Numerous electronic and electrical devices connect with the help of the Internet and support a smart city. Smart monitoring devices are used, such as advanced metering infrastructure, the IoT and so on. As a result, hardware and technology enable us to be more intelligent and make various features of smart cities relevant and accessible. A complete assessment of the smart city concept and its many uses, benefits, and preferences is provided. In addition, many potential IoT technologies are discussed with their applications. The usage of technologies to enhance innovation is a topic that is now generating a lot of discussions. Meanwhile, certain actual experiences from all around the world have been gathered, and the major roadblocks to its use have been removed. Figure 12.1 represents smart city–based linkages.

12.1.4 IoT TECHNOLOGIES FOR SMART CITIES

The IoT is a system which uses a conventional protocol to communicate, with the Internet serving as the connecting point. The IoT uses the critical notion of the

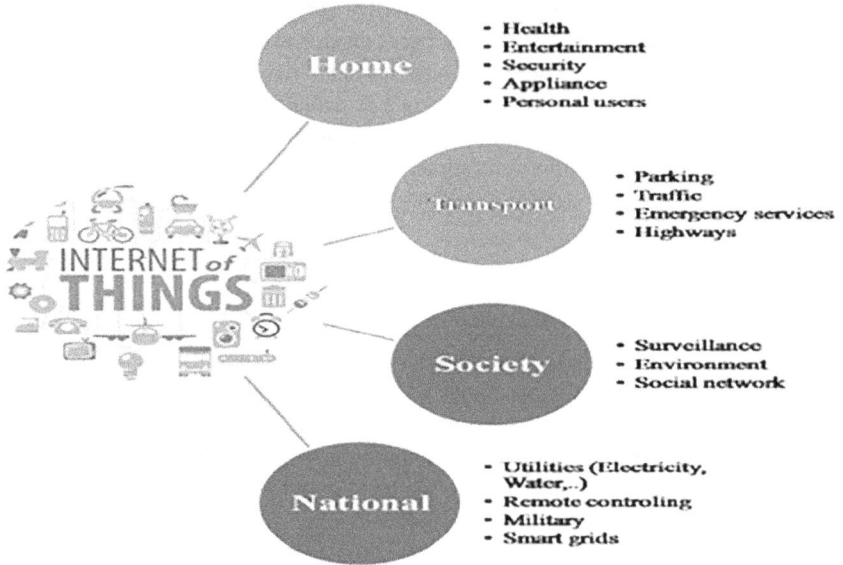

FIGURE 12.1 IoT-based linkages.

pervasive existence of objects that can be measured and inferred; it also changes the scenario. As a result, IoT expansions are aided by various factors, including communication equipment. Smart hardware, such as mobile phones, and other amenities, such as appliances, landmarks and foodstuffs, are examples of things in the IoT that may work together to achieve the aim.

12.1.5 IoT Application for Smart Cities

As a concept, IoT uses the Internet to connect disparate objects. Since being granted access to the Internet, everything that exists must be connected. These cities are connected through network sensors, and the appliances are connected intelligently and monitored remotely for their operations. Energy use can be managed by efficiently managing operations of air conditioners, lighting and so on.

12.2 LITERATURE REVIEW

According to [1], the smart city is a relatively new concept because a well-run city benefits both the government and the private sector. IoT connects billions of sensors to the Internet to be a productive and viable asset across the board in smart cities. An organization presented a study on the smart city concept based on IoT and big data breakthroughs, as well as an overview of smart city development in Romania and four smart city national agreements. A smart city uses computerized improvements or data and communication technologies (ICT) which improve the quality and efficiency of urban service administrations, improving city residents' lives. For visitors, it becomes easier to explore, and for locals, it becomes increasingly lucky. City and metropolitan governments have been given a far better way to interact with citizens and visitors. A city increasingly connected to its kin works and feels much better [2].

In a digital environment, smart urban areas play a significant role in developing novel administrations through modern innovation and outstanding arrangements. The IoT is a crucial enabler and motivator in densely populated places. In order to investigate the projected smart city administrations driven by IoT, we directed a special writing audit. We defined the examination questions, the search technique, selection criteria, information extraction quality, and quality assessment. What role does the IoT play in developing smart city administration? The articles were sorted and proposed or depicted in the smart city administrations. This research focused on conducting a thorough audit of active urban areas using IoT-enabled administrations. The audit revealed that this is a reasonably active research area, as seen by the growing number of papers published in the last few years. The findings have revealed that the IoT is the major enabling agent and driver of various smart city administrations. This facilitates the transformation of current city administrations into astute administrations as well as the establishment of new astute city administrations [3]. This study investigates the IoT-based smart city ecosystem and offers some recommendations. App developers, devices producers, and middleware developers are all classified as part of the user's ecosystem and stakeholders in this article. This chapter suggests various research issues as a proposed study direction for tackling the challenges faced by the smart city ecosystem, particularly in the coordination domain, which is a significant hurdle in creating sustainable IoT-based smart city ecosystems.

The architectural perspective can also be used to create a simulation model that can be parameterized to assess the ecosystem's performance [4]. According to ICT, a trend of outfitting common objects with care has emerged in recent decades intending to make human life more comfortable. Smart Cities worldview comes as a response to the intent of developing the city of the future, where citizens' prosperity and privileges are secured, and industry and urban planning are examined from a natural and economic perspective. Even though smart cities encounter some challenges in their implementation, more research activities in smart cities are being financed and carried out regularly. Furthermore, smart city features are being implemented worldwide to improve administrations or citizens' pleasure. Enabling sustainable mobility is considered one of the main aims of the smart city's creative and perceptive, according to [5], and the advancement of clever stopping frameworks speaks to a vital segment. The smart parking system (SPS) currently relies on various technologies, including WSN, RFID, cloud, NFC and mobile. It can aggregate ecological data, measure the inhabitance status of halting areas in real time, and route cars to the closest empty parking spot using a specific created program application. For this purpose, an NFC-based contactless e-wallet device is used to allow customers to pay for stop expenses. In addition, a redesigned programming program application would be installed on the cloud platform to monitor prepared exercises. In this case, the traffic cop is immediately alerted, and the traffic cop, with the help of an Android mobile application designed specifically for the situation [6]. This paper highlights the fundamentals of IoT, provides one-of-a-kind comprehensions, introduces IoT records, exhibits important IoT technology, and describes IoT applications.

According to the EU, the IoT is not always but matters, with only a few promising options for the generation projected to significantly impact our society after 5 to 15 years. The IoT market in China is also expanding. In the next few years, it will be a trend to drive economic growth by attracting large investments in affiliated sectors. However, China continues to create the same old IoT; the applicable generation is still in its infancy. However, the IoT will change our lives and make us much smarter.

Since the inception of Bitcoin in 2008, blockchain technology is emerging as the next progressive invention [7]. Even though blockchain started as a Bitcoin core invention, its applications are expanding to include accounts, the IoT, security, and other areas. Currently, a large number of private and public domains are embracing innovation. Aside from that, we'll see the birth of IoT as programming and equipment develop. Furthermore, those IoT devices must communicate and coordinate with one another. However, in situations with dozens or millions of IoT devices connected, we predict that using the present server-customer paradigm will have some limitations and challenges during synchronization. In this fashion, we propose that blockchain be used to build an IoT framework. IoT devices can be controlled and developed using blockchain [8]. This article discusses the current state of IoT development in China, including intentions for research and development, applications and standards. From China's perspective, this study depicts such challenges in terms of advancements, applications, and institutionalization and then suggests a three-stage open and generic IoT design to address the engineering challenge. Finally, this study discusses the opportunity and potential of IoT [9]. IoT is infiltrating many businesses' day-to-day operations. Smart homes, smart cities, smart frameworks and physical security, e-wellbeing, board access, and co-appointments are examples of applications. Sharp urban regions, for example, are forming in various territories, with redesigned street lighting controls, system checks, open prosperity and perception, physical security, shot ID, meter scrutinizing, and transportation assessment and streamlining structures being implemented on a city-wide scale [10].

Smart cities are interconnected systems or systems of systems that are complex. As a result, the smart city board is significant due to the vast number of partners, a good range of use spaces, the heterogeneity of data sources, and the complexity of smart frameworks. In contrast to other specialist approaches, less research has focused on the administrative aspect of smart cities. When considering the many stages of smart city development, a holistic vision is required to oversee the various stages of the smart city lifecycle. This paper presents a two-part framework for smart city lifecycle management. First and foremost, the lifecycle-based depiction of vibrant urban communities includes the profundity of "time" as another metric. Second, there is a link between the lifecycle of the board framework and the stages of IoT. As a result, using a limited case-based study, this paper examines the observational pertinence of lifecycle executives to smart city improvement [11]. Smart cities are unpredictably complex ecosystems made up of interconnected frameworks or assemblages of frameworks. Because of the large number of partners, diverse utilization spaces, heterogeneity of information sources, and complexity of smart frameworks, the board of smart cities is substantial. Nonetheless, in contrast to other specialist perspectives, investigations have focused on the administrative aspect of smart cities. When considering the many stages of city development, a comprehensive vision is required to manage a magnificent city throughout its existence. This research presents a two-part methodology based on the brilliant city lifecycle. To begin, a lifecycle-based depiction of intelligent smart communities includes the depth of "time" as a measurement. Second, a collaborative approach between the board structure and the phases of IoT. Smart cities have drawn in global logic and business consideration, and specialty advertising is being established, which connects almost all business segments [12]. Local governments have moved toward the smart city set in an attempt to enable and focus on habitats, visitors, and initiatives. However, capturing the smart city setting cannot be free, and comparing speculations is a vast and high risk without suitable management. Furthermore, investing in the smart city sector does not guarantee that vital connections will be made, and governments and sellers will require increasingly effective tools. This document is a work in progress that examines innovative city action strategies. While an extensive plan of action portfolio is suitable for keen municipal partners, displaying can distinguish where related advantage originates and how it streams. Emerging technologies can also have a larger impact on various aspects of the smart city [13].

12.3   RESEARCH METHODOLOGY

Research design: An exploratory and descriptive research design has been followed in this research. The literature review has given many insights into the factors to be considered in the present study. For the descriptive research, non-probabilistic sampling has been used. Various variables to be considered for the study have been extracted from the literature review; thereafter, the primary data was collected through a structured questionnaire. The data was collected from 1500 respondents living in smart cities. The initial questionnaire was pre-tested with 50 respondents staying in smart cities across India. Out of the total sample of 1500, 43 were found to be outliers. So, the final sample size was 1457.

This sample includes 71% males and 29% females. The age group distribution of the respondents was as follows: 48% below age 30 years, 29% between ages 31 and 45 years, 12% between ages 46 and 60 years, and 11% over age 61 years. Most of the respondents were from the service class (44%); 15.6% were from the business class, 32% were students, and 8.4% were retired. As to the income distribution of the respondents, 64.4% were from the income group of Rs. 1 to 7 lakhs per annum; 17.8% were from Rs. 7 to 10 lakhs per annum; 6.2% were from Rs. 10 to 12 lakhs per annum; and 11.6% were above 12 lakhs per annum. Of the sample, 78.9% believed that the development of smart cities improves people's lives, and 15.8% believe it does not; the remaining 5.4% do not have any clear thoughts regarding the role of smart cities. Most (60.5%) agree that building a fully energetically and ecologically autonomous smart city will be very useful. The rest, 39.5%, don't agree on this same basis. 76.3%. Agreed, and the rest, 15.8%, have agreed but are not sure the economic cost will be worth it for a smart city. Although a percentage of a person, 7.9%, who have not agreed to economic cost will be worth it for smart cities. 78.9%. Agreed, and the rest 18.4% have also agreed, but according to them, being Smart is not essential for the future.

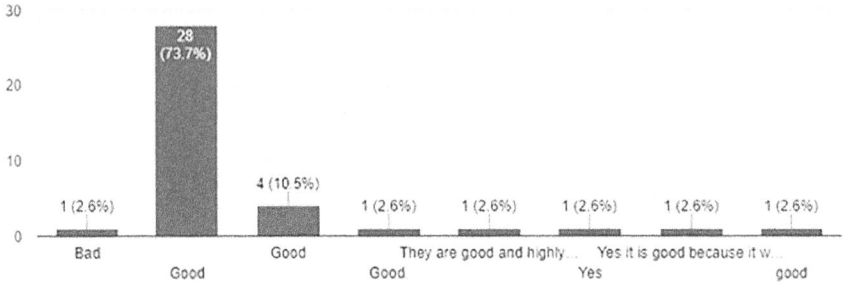

FIGURE 12.2 Survey: perception of smart city.

As per the survey (Figure 12.2), 73.7% of people agreed they think smart cities are broadly good, and the rest, 23.5%, also agreed with some other parameter related to smart city. However, as per the survey, smart cities are broadly good and highly appreciative, and 2.6% of people do not agree that smart cities are broadly good. It is also observed that the majority of people (92.1%) believe that smart cities exist, and the rest (7.9%) believe that they do not.

12.4 DATA ANALYSIS

The purpose of the presentation is the close relationship between consumer preferences and efforts to provide cloud computing companies. The representative tool (regression analysis) was used to shore it to the best instrument. Analyzing the moving pressure to link the change of care, we try to explain (the preference to move from one place) to one or more variables to be granted.

In statistics, the coefficient determination means R^2 ("R squared"), which is utilized in statistical models as the context whose principal reason for existing is related to future results based on other data. R^2 is observed as a number between 0 and 1, used to depict how well a relapse line fits much information (Table 12.1).

An R^2 of almost 1 indicates that a regression line fits the information well, while an R^2 near 0 shows that a regression line does not fit the information quite well. The statistical model accounts for the proportion of fluctuation in a dataset. It provides a future outcome of how well the model will likely predict measures. The present research of $R = 1$, which is equal to 100%, means that the independent factors clarify 100% of the variety in the dependent variable (Table 12.2). The worth is high, maybe because of the smaller example size, yet at the same time it is not zero, which implies that the model clarifies the variety in the model.

TABLE 12.1
Dependent Variable: Consumer Preference

Regression Statistics	Multiple R	R Square	Adjusted R Square	Standard Error	Observations
	1	1	1	1.2313E-15	6

TABLE 12.2
Dependent Variable: Preference. Predictors: Constant

ANOVA

	df	SS	MS	F	Significance F
Regression	2	809.3333333	404.6666667	2.66914E+32	4.21288E-49
Residual	3	4.54828E-30	1.51609E-30		
Total	5	809.3333333			

df "the degrees of freedom in the source."
SS "the sum of squares due to the source."
MS "the mean sum of squares due to the source."
F "the *F*-statistic."

TABLE 12.3
Dependent Variable: Preference

	Coefficients	Standard Error	t Stat	P-value	Lower 95%	Upper 95%	Lower 95.0%	Upper 95.0%
Intercept	100	1.62141E-15	6.16748E+16	9.40044E-51	100	100	100	100
No	−1	2.2115E-16	−4.52181E+15	2.38525E-47	−1	−1	−1	−1
May be	−1	4.8321E-17	−2.06949E+16	2.48816E-49	−1	−1	−1	−1

There is a relationship between IoT products for smart cities and the following factors by their respective magnitude, as shown in Table 12.3; the coefficients are represented by their respective percentages. A direct relationship exists between dependent (Preference) and independent variables. This means the significance level of the overall regression model is greater than 0.05, so it does not come under the acceptance region. So our regression model is valid, and the variance of a dependent variable (Preference) is explained by independent variables, which include consumer perception about the following: IoT products for smart cities are necessary future platforms for the upgrading of various business modules as per customer preference.

From Table 12.3, it can be clearly observed that all the B coefficients are not all positive, which indicates that there is not a high association between consumer preference and perception of smart cities of IoT products for sales and development.

12.5   CONCLUSION

The significance of considering how the ideas are new for new technologies (particularly to the IoT) advantage is unquestionable for smart cities. The point of this project was to explore variation details and highlights of IoT frameworks, alongside the beneficial and powerful motivating forces for utilizing them. The achievement of the IoT substructures can enable a volume of chances for smart cities. The most noteworthy research inspirations were later communicated briefly, and a few primary and accommodating applications were clarified. The project sketched out how to step-by-step exercises could be broadened and improved through using them. Likewise, the challenges emerging from executing the IoT systems were sketched out. Across the board, use is one of the fascinating future inclinations in consolidating the IoT stage into other autonomous and smart systems to give intelligence through the IoT. Its functionality should be specified and employed with smart city systems and sensors to ensure residents' rights. Also, providing a strategy to adapt to some significant difficulties, for example, security privileges of the clients/residents, is still a region of research interest. Some of the advancements in smart cities for actual usage worldwide were presented, which can be considered as the sample test for future comprehensive smart cities.

REFERENCES

[1]. Rotuna, C., Smada, D., & Gheorghiè, A. (2017). Smart city applications built on big data technologies and secure IoT. *Ecoforum Journal*, *6*(3).

[2]. Mijac, M., Androcec, D., & Picek, R. (2017). Smart city services driven by IoT: A systematic review. *Journal of Economic and Social Development*, *4*(2), 40–50.

[3]. Oktorini, R., & Barus, L. S. (2022). Integration of Public Transportation in Smart Transportation System (Smart Transportation System) in Jakarta. *Konfrontasi: Jurnal Kultural, Ekonomi Dan Perubahan Sosial*, *9*(2), 341–347.

[4]. Sánchez-Corcuera, R., Nuñez-Marcos, A., Sesma-Solance, J., Bilbao-Jayo, A., Mulero, R., Zulaika, U., . . . & Almeida, A. (2019). Smart cities survey: Technologies, application domains and challenges for the cities of the future. *International Journal of Distributed Sensor Networks*, *15*(6), 1550147719853984.

[5]. Mainetti, L., Patrono, L., Stefanizzi, M. L., & Vergallo, R. (2015, December). A Smart Parking System based on IoT protocols and emerging enabling technologies. In *2015 IEEE 2nd World Forum on Internet of Things (WF-IoT)* (pp. 764–769). IEEE.

[6]. Liu, T., & Lu, D. (2012, August). The application and development of IoT. In *2012 International Symposium on Information Technologies in Medicine and Education* (Vol. 2, pp. 991–994). IEEE.

[7]. Huh, S., Cho, S., & Kim, S. (2017, February). Managing IoT devices using blockchain platform. In *2017 19th International Conference on Advanced Communication Technology (ICACT)* (pp. 464–467). IEEE.

[8]. Chen, S., Xu, H., Liu, D., Hu, B., & Wang, H. (2014). A vision of IoT: Applications, challenges, and opportunities with China perspective. *IEEE Internet of Things Journal*, *1*(4), 349–359.

[9]. Centenaro, M., Vangelista, L., Zanella, A., & Zorzi, M. (2016). Long-range communications in unlicensed bands: The rising stars in the IoT and smart city scenarios. *IEEE Wireless Communications*, *23*(5), 60–67.

[10]. Hefnawy, A., Bouras, A., & Cherifi, C., 2018. Relevance of lifecycle management to smart city development. *International Journal of Product Development*, *22*(5), pp. 351–376.

[11]. Hefnawy, A., Bouras, A., & Cherifi, C. (2018, July). Does end of life matter in smart cities?. In *IFIP International Conference on Product Lifecycle Management* (pp. 442–452). Springer, Cham.

[12]. Anthopoulos, L. G., & Tsoukalas, I. A. (2006). The implementation model of a Digital City. The case study of the Digital City of Trikala, Greece: e-Trikala. *Journal of e-Government*, *2*(2), 91–109.

[13]. Hansain, S., Gaur, D., & Shukla, V. K. (2021). Impact of emerging technologies on future mobility in smart cities by 2030. *2021 9th International Conference on Reliability, Infocom Technologies and Optimization (Trends and Future Directions) (ICRITO)*, pp. 1–8. doi:10.1109/ICRITO51393.2021.9596095.

13 Smart Intelligent Approaches for Healthcare Management

Simra Nazim, Vinod Kumar Shukla, Fatima Beena and Suchi Dubey

13.1 INTRODUCTION

Artificial intelligence (AI), machine learning (ML) and deep learning (DL) have been used interchangeably in technology [1–2]. AI is a broad area in computers, and ML and DL come under the AI umbrella. ML, a subset of AI, will help medical professionals to take decisions more accurately, as the machines can learn from the available data [3–4]. This will help machines look into every aspect of the past data and make better decisions that suit the current problem. The models are trained to make predictions and help detect disease in advance detection. Diagnosis, drug discovery and manufacture, tailored medication and treatment, smart hospital records and advances in computer vision have all contributed to the development of medical imaging using ML [5–6]. The DL approach, a subset of ML techniques, attempts to

DOI: 10.1201/9781003218715-17

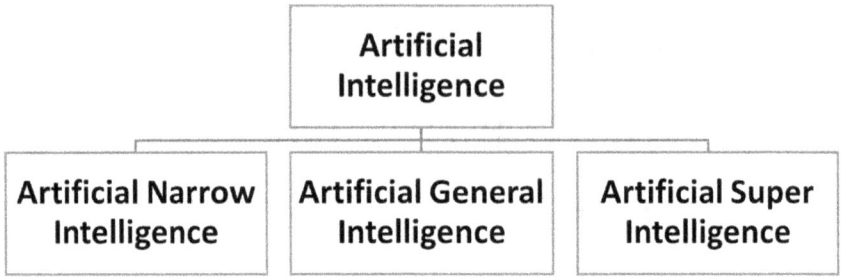

FIGURE 13.1 Types of artificial intelligence.

mimic the human brain. It uses a neural network with three or more layers and can be supervised, semi-supervised or unsupervised. It analyzes data at great speeds without any compromise on accuracy.

13.1.1 ARTIFICIAL INTELLIGENCE

AI is intelligence demonstrated by machines, making them smarter and enabling them to think without human intervention. AI can be categorized into three types (Figure 13.1).

Artificial narrow intelligence (ANI) is a sort of artificial intelligence that is target oriented and created to carry out a particular task. Artificial general intelligence (AGI) is a technology that allows machines to learn, perceive and act in a manner that is barely distinguishable from humans in a specific circumstance. Regarding intellect, ASI (artificial super intelligence) is a theoretical AI in which machines can surpass even the most intelligent humans.

13.1.2 MACHINE LEARNING

ML, a subfield of AI, tries to create intelligent systems through statistical learning approaches. ML systems can learn and develop independently without being explicitly programmed. In our everyday routines, we use ML-based systems to recommend video and music online streaming based on our preferences and previous viewings.

13.1.3 DEEP LEARNING

DL is a branch of AI built on how the human mind absorbs knowledge and improves from past experiences. DL methods in predicting and classifying data enhance a computer model's ability to filter incoming data through layers. When it comes to analyzing data, DL is comparable to the human brain. Different types of DL network models include CNN (convolutional neural networks), long short-term memory networks (LSTMs), and multilayer perceptron (MLP). It has been used in self-driving automobiles, virtual assistants and others.

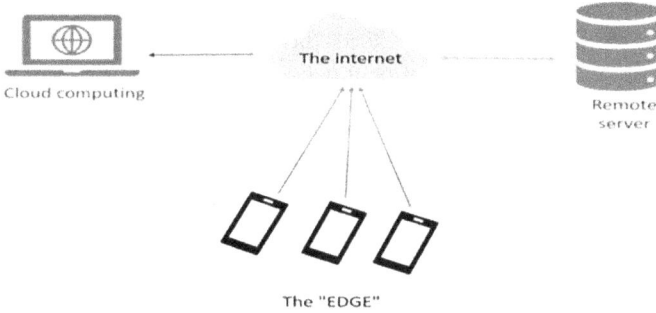

FIGURE 13.2 Edge AI and its various configurations.

AI and ML have advanced quickly in recent years. Medical image processing, computer-aided diagnosis, image interpretation, image fusion, image registration, image segmentation, image-guided therapy, image retrieval and analysis have all benefited from ML and AI techniques, which we will discuss in more detail in the following sections [7].

13.1.4 EDGE AI

With the expansion of the Internet of Things (IoT) in the healthcare system, the number of devices connected is also increasing. The devices collect data in real time and require more storage and computational capacity. Figure 13.2 shows the common IoT system configuration with ML applied in today's workspace.

All medical sensors and devices must be connected to the Internet through a common device, which sends raw data to the backend server. The servers use ML to help analyze and predict to make use of the data collected. The problem begins when the devices clog the network traffic, and data might not be transferred on time because of traffic on the local Wi-Fi used by the hospital. The limitations of AI in the cloud environment can be summarized as follows:

Connection to the Internet: Remote locations can have difficulty accessing the Internet, and thus utilizing AI for medical emergencies would be interrupted.
Transferring data to the cloud: This is a task as medical images need to be as detailed as possible; thus, transferring and maintaining such large files increases the cost for medical organizations.
Delays: The images can take time to be transferred to the cloud, which would take much time for the doctors to wait for those images to be analyzed.

Applying AI algorithms on devices or local servers is known as edge AI, a possible solution for the above problems. The technology focuses on using neural networks on end devices, also known as edge devices, the computing process is done on device instead of in the cloud. Edge computing tries to connect numerous edge devices and servers in close range to process a huge amount of data. At the same time, AI aims to imitate human behavior and thought processes by learning the data. Thus, combining the two and pushing AI to the edge extends the number of benefits. Edge intelligence does not imply that the model will be trained or interpreted into the edge devices. However, an edge device-cloud cooperation mode of operation can be used where only relevant data is uploaded to the cloud for further processing and data that can be processed on-site will be saved offline.

13.2 EDGE AI AND MACHINE LEARNING IN HEALTHCARE

13.2.1 EDGE AI FOR HEALTHCARE

AI can help doctors and medical staff with tedious tasks by automating the workflow in administrative work, which can reduce the working hours spent on repetitive tasks and cut down costs. Many of the applications are natural language processing (NLP)-based chatbots and robotic process automation. Electronic records for health systems using AI is an upcoming application that can define the next-generation medical system powered by AI. Edge AI combines edge computing, IoT and AI, bringing the best technology. This technology can help healthcare systems create smart hospitals that connect and communicate seamlessly while taking care of patients on edge.

For example, we have Internet-connected sensors in hospital rooms that send a multitude of data to servers across the Internet. We can implement ML models to learn and predict outputs. All these useful statistics can be shared with doctors in their respective remote dashboards, where they can determine what actions to take next from the comfort of their offices. This is cloud computing in healthcare. Once the data increases and we scale up operations, we might run into physical limitations in the network bandwidth. Wi-Fi might get crowded with all the sensors, or we might have to pay more for Internet service providers for extra bandwidth. This is where edge computing is introduced, where you can run local computers or servers to help manage all that data. These servers do not exist in the cloud. You likely own or have control of them, so they are on the edge of the cloud. They may not be as powerful as the remote servers, but they can help alleviate some bandwidth. These servers can collect, organize and can do some basic analysis of the data before shipping it off to the remote server. Once ML algorithms are running on that local server, we refer to the system as edge AI. They will be more helpful and closer to the collection devices rather than relying on the Internet and its latency. When ML algorithms on the collection devices assume the processing power is there, we could do some basic data analysis and curation before sending it off to our servers.

Edge AI has critical benefits in healthcare, as follows:

Data privacy: We can process data locally on the device itself, allowing us to avoid uploading data to the cloud for processing. This is a huge advantage as it can protect our data from hacks and cybertheft and make the users worry-free about how their data is being handled once uploaded [8].

Cost reduction: Since all the data is not uploaded to the cloud, it saves a huge amount of bandwidth and storage and cuts down processing costs. Processing medical data can be very expensive as the data collected is in huge amounts—gigabytes and petabytes per day—and edge AI can help reduce costs for medical organizations by uploading only relevant data for processing.

Low latency for offline execution: Real-time decision-making can face an issue if there is a network or connectivity failure. In the medical field, this can be risky, such as robots monitoring elderly people or those working at factory assembly lines. Deploying the AI models on edge devices can help run systems offline almost in real time, especially in remote areas.

Thus, edge AI helps the ML models to be deployed on devices with low memory and computational resources allowing an offline execution. The following section describes what AI brings to the healthcare industry.

13.2.2 Machine Learning for Healthcare

Over the past years, ML and AI have made a breakthrough in every industry. These technologies have benefitted the healthcare industry, especially the medical imaging departments, from image processing, segmentation, interpretation, registration and image-guided therapies to retrieval and analysis of image data. ML algorithms help gather information and make effective and efficient use of it by analysis. Doctors can use these ML algorithms to diagnose diseases and predict illnesses accurately to prevent them. This strategy can assess how changes in the body lead to certain diseases, helping researchers make conclusions more efficiently.

Techniques like support vector machines (SVMs) [9], neural networks (NNs), and k-nearest neighbor (KNN) are non-learning algorithms that are limited in their ability to interpret natural photos in raw form, can be tedious, take more time and rely on specialist knowledge. DL algorithms include popular techniques like convolutional neural networks (CNNs), generative adversarial networks (GANs), multilayer perceptrons (MLPs) and others, which are provided with raw forms of data which are learned quickly and automatically. The techniques learn the different levels of representation and data from the large datasets available to meet the desired outputs.

Even though autonomous disease identification based on traditional approaches in radiology has shown great precision for years, emerging developments in ML algorithms have sparked a DL revolution. DL-based algorithms exhibited improved accuracy and scalability in various disciplines, such as audio recognition, language processing, lip-reading, computer-aided diagnostics, target detection and clinical trials. The following are different types of ML approaches:

Unsupervised learning: Machine learning approaches that use unlabeled data are known as unsupervised learning methods. They are capable of identifying hidden layers or patterns in datasets with the help of humans and give useful insight into data that cannot be predicted by humans easily. Unsupervised learning techniques include clustering data points using a similarity metric and dimensionality reduction to project high-dimensional data into lower-dimensional subspaces (also known as feature selection). These approaches in healthcare have been used to predict heart disease using clustering and to predict hepatitis using principal component analysis (PCA), a dimensionality reduction methodology.

Supervised learning: The use of labeled training data to establish or trace the connection between inputs and outputs is known as supervised learning. If the result is discrete, the task is called classification; if the output is continuous, it is called regression. Two well-known examples of supervised learning approaches in healthcare are the classification of various lung illnesses and the identification of various body organs from medical imaging. ML approaches cannot be supervised or unsupervised when the training data contains both labeled and unlabeled samples.

Semi-supervised learning: Semi-supervised techniques are useful when both labeled and unlabeled samples are accessible for training a model, such as a small amount of labeled data and a substantial majority of unlabeled data. Because acquiring enough labeled data for model training in healthcare is complex, semi-supervised learning algorithms can benefit various healthcare applications.

Reinforcement learning: Reinforcement learning (RL) is a set of approaches for learning a policy function from data, actions, and incentives over time. One recent example is using RL models for context-aware symptom screening for disease diagnosis. Furthermore, a good example of the game of go, which uses RL and the fusion of supervised and unsupervised learning methodologies, beat a human champion player, showing the significance of implementing RL in healthcare.

Recent progress in ML and intense learning contribute to identifying, classifying and quantifying patterns in medical data. The capacity to use hierarchical feature representations acquired simply from data rather than features generated by hand based on domain-specific information lies at the core of these advancements. DL is gradually becoming the industry standard, resulting in enhanced outcomes and performance in medical applications. From the above description, we can conclude that by using different algorithms, ML can help healthcare in the following activities:

Classification: ML algorithms can help doctors classify and label diseases and special medical cases.

Clustering: Clustering algorithms are typically effective in predicting diseases by dividing similar patient data based on important attributes. So far, a large number of clustering algorithms have been designed to evaluate a variety of healthcare datasets.

Prediction: Utilizing available data and examining common attributes, ML algorithms can help predict medical conditions, focusing on making a prognosis on how a patient's current condition will evolve and preparing patients and doctors for required medications.

Recommendations: Based on available data, ML can provide us with useful recommendations without needing to search for them manually.

Anomaly detection: Helps to put out more relevant information and decide whether any action is required immediately.

Ranking: Arranges relevant information in order of priority.

Automation: Helps to automate repetitive tasks like scheduling meetings and appointments, data entry and so on, which saves time for doctors and nurses.

The technology helps doctors to focus on patients and their needs rather than search for and enter relevant data. It also increases diagnostic accuracy and develops a more precise medical plan according to their needs.

13.3 ROBOTIC SURGERY

Surgeries require accuracy and a steady approach for a long time with the ability to adapt to changing circumstances. Robots can plan surgeries, simplify tasks and help surgeons with long surgeries. Still, in its early stages of development, robotic surgery will become a new revolution in using ML in healthcare. The following are a few advantages of robotic surgery:

Minimally invasive surgery: Instead of utilizing the surgeon's hands, robotic surgery involves reaching inside the patient's body using very small equipment. In most cases, incisions can be as small as a centimeter in length and are referred to as "keyhole" incisions. With a camera installed on one of the surgical robot's arms, the robot can navigate weak veins, reducing the risk of blood loss. While large incisions are usually necessary for difficult-to-access sections of the body so that the surgeon may examine the area, robotic surgery allows the surgeon to maneuver around obstacles simply.

Reduced recovery time after surgery: When surgery is minimally invasive, the body heals faster. The healing time is reduced because the incisions are smaller using robotic surgery. Patients who have undergone minimally invasive procedures can usually go home after a few days, and shorter hospital stays are frequently associated with a quicker recovery. Less bodily injury and fewer, faster-healing scars result from smaller incisions.

Infection risks are reduced: Infection is one of the most important concerns for individuals who have had surgery. When large portions of the body are opened, the risk of infection increases, which could make recovery time longer and result in a lengthier stay in the hospital. Because robotic surgery is minimally invasive, the risk of infection and the associated adverse effects are reduced.

Clinically improved results: Robotic surgery, as compared to traditional surgery, can help reduce errors during complicated surgeries. When the surgical team has great precision and a detailed vision of the operating region, the danger of harming tissue is reduced, and the surgery is more likely to be successful. Table 13.1 lists AI models implemented in robotic surgery [10].

13.4 PREDICTIONS FOR INFECTIOUS DISEASE OUTBREAKS

Infectious illness occurs when a person is infected by a virus (such as COVID-19) from another person or animal. It causes harm not only to individuals but also to society as a whole and is thus classified as a societal issue [18]. Communicable

TABLE 13.1
Artificial Intelligence Models Implemented in Robotic Surgery

Paper	Outcome	Source of Data	Main Conclusion
Automated robot assisted surgical skill evaluation: predictive analytics approach [11].	Tool for evaluating skills	The robotic arms' overall movement characteristics are as follows: • Time to complete the task • Perception of depth • Smoothness of motion • Curvature • Angle of rotation	Expert and inexperienced robotic surgeons are automatically distinguished.
Biaxial sensing suture breakage warning system for robotic surgery [12].	Haptic feedback surrogate	Sensor on robotic instruments	Suture breaking is prevented by using a haptic feedback device, which also improves knot quality and learning.
Utilizing machine learning and automated performance metrics to evaluate robot-assisted radical prostatectomy performance and predict outcomes [13].	Postoperative outcomes	Metrics of performance that are automatically generated: • Total time spent on completion • Total time spent idle • The total length of all instruments' paths • Dominant/non-dominant instrument path length, moving time, mean velocity, and idle time • Adjusting the camera's position, frequency, path length, moving time, mean velocity, and idle time • Energy consumption • Replacement of the third arm • The use of a clutch	A machine learning system is being used to estimate the length of a hospital stay and the duration of a Foley catheter.

A deep learning model using automated performance metrics and clinical features to predict urinary continence recovery after robot-assisted radical prostatectomy [14].	Postoperative outcomes	Metrics of performance that are automatically generated: • Metrics that are time related • Kinematic metrics from instruments • Metrics for camera movement • Metrics for system events • Characteristics of the patient: age, body weight, and body mass index • Gleason score before surgery • ASA Classification • Operative time -Lymphadenectomy extent	Performance indicators are more relevant than patient characteristics in predicting continence following robotic radical prostatectomy.
A machine learning approach to predicting case duration for robot-assisted surgery [15].	Intraoperative outcomes	Variables used in model construction: • Procedure scheduled • A procedure group is a collection of people who come up with new ways to do things. • Age-obesity-gender-combined situation Model of a robot • Malignancy • The location of the tumor • Hypertension • History of smoking • Atrial fibrillation is a condition in which the heart beats irregularly • Coronary artery disease • Diabetes • Cirrhosis • Diabetes • Anesthesia provider • Surgical assistant	Higher accuracy for the prediction of case duration

(Continued)

TABLE 13.1
Continued

		• Month of the year • Time of day • Day of the week	
Deep learning with convolutional neural network for objective skill evaluation in robot-assisted surgery [16].	Tool for evaluating skills	Data from the manipulator of the master tool and the manipulator on the patient's side.	Automatic assessment and feedback of surgical abilities using an artificial neural network.
Automatic and near real-time stylistic behavior assessment in robotic surgery [17].	Tool for evaluating skills	Position sensors on the surgeon's limbs Stylistic behavior metrics: —Fluid/viscous —Smooth/rough —Crisp/jittery —Swift/sluggish —Calm/anxious —Relaxed/tense —Deliberate/wavering —Coordinated/uncoordinated	The quality of stylistic behavior is used to assess surgical skill.

disease tracking is a complex procedure wherein data on emerging infectious diseases and carriers is gathered, evaluated and analyzed in a structured and continuing method [19]. Furthermore, the findings are rapidly disseminated to those requiring them to prevent and control infectious diseases. The COVID-19 outbreak shows our unpreparedness towards communicable diseases and how they can affect the entire world in just a few months [20–22]. ML- and DL-based systems can help predict infectious diseases early, preparing everyone to be safe and avoid the global effects of this virus. Some of the notable ML algorithms that have helped detect infectious diseases are shown in Table 13.2 [23].

This research [23] examines ML and soft computing algorithms for predicting the COVID-19 outbreak as an alternative to SIR (susceptible, infected and recovered) and SEIR (susceptible, exposed and infected, but not yet infectious), infectious (now can infect others) and removed/recovered) models. Out of many ML models studied, two (multi-layered perceptron [MLP] and adaptive network-based fuzzy inference system [ANFIS]) showed promise. This work suggests that ML could be a viable technique for simulating the COVID-19 outbreak, based on the findings and given the exceedingly complicated nature of the outbreak and its variation in behavior from country to country. The project is a preliminary measuring exercise to demonstrate how ML could be applied in future studies. It also suggests that the true novelty in outbreak prediction may be accomplished by combining ML with SEIR models.

13.4.1 CLINICAL RESEARCH

AI systems can be used to automatically analyze clinical trial electronic criteria records and connect them with clinical trial recruitment from trial releases, social media and databases [21]. Patients will have a chance to become more aware of clinical trials they are interested in. They can approach the researcher earlier and attempt assessments and evaluations. The use of AI in clinical trials has led to about a 58.4% rise in attendance at lung cancer trials [30].

Incorporating AI approaches with wearable technology allows for efficient, real-time, and tailored patient monitoring, which can be done automatically and constantly

---

TABLE 13.2
ML Algorithms for Outbreak Prediction

Authors	Infection Outbreak	ML Algorithm
Tapak, L.; Hamidi, O.; Fathian, M.; Karami, M. [24]	Influenza	Random forest
Koike, F.; Morimoto, N. [25]	H1N1 flu	Neural network
Anno, S.; Hara, T.; Kai, H.; Lee, M.A.; Chang, Y.; Oyoshi, K.; Mizukami, Y.; Tadono, T. [26]	Dengue fever	Neural network
Chenar, S.S.; Deng, Z. [27]	Oyster norovirus	Neural network
Liang, R.; Lu, Y.; Qu, X.; Su, Q.; Li, C.; Xia, S.; Liu, Y.; Zhang, Q.; Cao, X.; Chen, Q., et al. [28]	Swine fever	Random forest
Raja, D.B.; Mallol, R.; Ting, C.Y.; Kamaludin, F.; Ahmad, R.; Ismail, S.; Jayaraj, V.J.; Sundram, B.M. [29]	Dengue/Aedes	Bayesian network

throughout the trial. This can help ensure that protocol requirements are followed, and those endpoint assessments are accurate. DL models can build patient-specific illness diaries tailored to changes in behavior and disease symptoms by accessing data through sensing devices and monitoring. Such dynamic illness diaries make collecting compliance and goals more reliable and efficient. Image-based endpoint detection would benefit from ML algorithms that have been authorized to identify medical images. In simulated trials, ML-based algorithms were used to estimate the smallest and fewest doses required to decrease brain tumors while reducing chemotherapy side effects. This could lower the number of students who drop out due to safety concerns.

13.4.2 ELDERLY CARE

Elderly people face everyday challenges, and most people are unaware of their difficulties. People do not prefer to live a dependent life, and making their lives feel more independent with AI can make their lives more comfortable [31]. Using analytics and ML algorithms helps approach a predictive healthcare system. Implementing smart home systems can help them live at home comfortably, so mobility patterns and behavioral changes can be studied easily, leading to more accurate healthcare predictions [32–34].

13.4.3 HELP PEOPLE WITH SPECIAL NEEDS

AI can help people with disabilities by providing personalized solutions and models focused on their special needs. People with disabilities now do not have to depend on others completely and can utilize AI to simplify tasks in their daily life. Nowadays, many platforms and apps are available, and wearables help people do simple tasks [35–36].

For example, image recognition using DL can help identify and recognize objects and people around people who have visual impairments. This can be combined with speech recognition for commands and a text-to-speech system that can recognize the people around them and say it to them. It can help them avoid obstacles in their daily tasks. Lip-reading recognition systems can help people with hearing disabilities. Optical character recognition using ML/DL can help people read out the text around them.

These technologies can help people with special needs to connect and communicate with everyone with minimum help, access the same services as normal people, move around and live independently.

Virtual assistants such as Alexa, Google Home, and Siri allow persons with limited mobility to use smartphones by speaking commands into their phones. The Braille AI tutor is a novel way to compensate for the scarcity of Braille teachers. Seeing AI for iOS is a visually impaired program that can read and explain any document held up to the camera on a smartphone, such as money or mail. It can even detect photos, colors, and people, enabling it to deduce information about people's emotions [37–38].

13.5   MEDICAL IMAGING

Imaging modalities are a technique of medical imaging which use internal patient signals that reflect either anatomical structures or physiological events are detected using a specific physical mechanism [39].

As we know, medical images consist of a huge range of modalities and a high level of pixel density. Numerous modalities are available, along with many being developed, such as spectral CT even with widely used imaging modalities, and the pixel density has improved with a rise in data density (Figure 13.3). Medical images are available in bulk, and due to patient privacy, they can be spread among many clinics. Given that there is no normalized standard for the equipment used in clinics, there is a drift in the images gathered. It is rare to find a centralized source for medical image big data. Patterns of disease are numerous, and these images' interpretability is hard. Along with this, the labels associated with the images are very noisy. Labeling these images is a time-consuming task. Each image would need annotation, which can be in different forms per the image's requirements. These tasks need to be fulfilled accurately and carefully is essential.

The samples are unbalanced and diverse: The existing labeled images appear different from one sample to another because their probability distribution exists in various diversities. The proportion of positive to negative samples is very unbalanced. For example, the pixel value in a particular illness, say a tumor, is often one too many orders of magnitude lower than that of normal tissue. Processing these images involves high complexity and a series of tasks and technologies from labeling, classification, enhancement, detection and registration of images. The tasks can become complex when combined with multiple diseases and modalities. The application of AI in medical imaging has been one of the most promising areas of innovative healthcare, from image

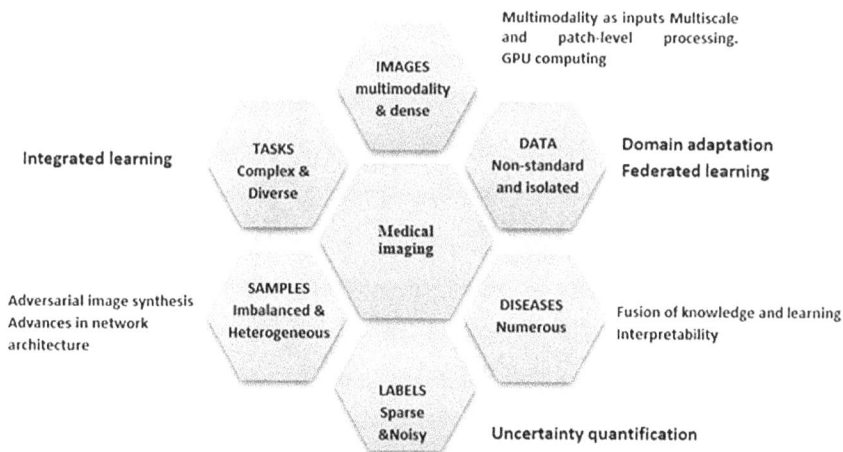

FIGURE 13.3   Medical imaging and its various modalities.

processing to detection and interpretation of data involved, and it will help radiologists' workflow.

In medical imaging, the more images and examinations involved, the better results are achieved. The radiologist needs to analyze images and evaluate contexts. But as the number of images increases, more time will be spent on the analysis, squeezing the time for clinical evaluation [40], and the interpretation of images would be given to another physician. This can cause a drift as the analysis and interpretation will be made by different physicians. This is threatening to both radiologists and patients as the interpretation of images might be handed over to non-radiologist who does not have expertise in medical imaging. Thus, AI can help overcome workflow issues in medical images to a great extent.

AI algorithms identify patterns by being trained on many images and examinations and give information about the characteristics of the findings. DL architectures have developed radionics models that focus on extracting quantitative features in radiology images. Undoubtedly, data from radiomics research, like intensity, form, texture, wavelength and so on can be derived from disease diagnosis and integrated into ML approaches, providing information for predicting treatment response, differentiating malignant or benign tumor cells, and evaluating cancer genetic traits in a wide range of cancer types [41].

By improving picture quality and lowering radiation dose, AI applications may enhance the accuracy of practical protocols, lowering X-ray scanning time [42] and optimizing personnel and CT/MRI scanner utilization, cutting costs [43]. These programs will make technicians' jobs easier and faster, resulting in greater technical inspection quality on average. This could help AI systems overcome one of their weaknesses, namely their inability to perceive the impacts of location, motion distortions and other factors exacerbated by the lack of standardized acquisition techniques. Put another way, AI requires high-quality research, but its implementation will improve quality. The golden age of radiological standards may be reached, boosting productivity. The following are a few popular ML approaches in medical imaging [44]:

Classification: Classification is a critical stage in medical imaging. ML systems get many images as input for processing, resulting in a usable output.

Localization: After the classification, identifying the location of the disease in the image accounts for defining the output positions. This is known as localization. It is a diagnostic pre-processing procedure that enables the radiologist to distinguish many aspects of medical images.

Detection: Detection helps use object recognition to distinguish one instance from another. Object identification is important in medical image processing, diagnostics and monitoring.

Segmentation: Image segmentation is crucial for illness identification in medical image processing. Splitting a visual image into numerous sections is known as image segmentation. Medical image segmentation aims to make it easier and more convenient to examine visual pictures. The result of segmentation in a medical image is a succession of medical segments that make up the entire medical image.

Registration: The process of transforming many datasets into a single coordinate model is known as registration. The registration of a picture is crucial in the domains of biological imaging and medical imaging. You must first register to evaluate or combine data from various medical sources. To reduce perceived overlap between photos, a medical technician is usually advised to view them differently.

Different types of algorithms can be implemented in medical imaging related to ML, such as the following:

KNN (k-nearest neighbor): In KNN [45], a set of features for a particular unknown sample object, known as input vectors, are classified by assigning it to the most similar class. In this case, the number of neighbors, or known objects nearest to the sample item, is k. These are the votes for the classes to which the sample object might belong. If k = 10, the unknown item is simply assigned to the closest neighbor's class. The similarity function, which determines how similar one example object is to another, will be the Euclidean distance between the values of the input vector and the values of the vector.

A paper [46] used ML on SI analysis of lymph node metastasis in gastric cancer by using metric learning to reduce feature space and then used the KNN algorithm as their classifier to differentiate and classify non-lymph node metastasis from lymph node metastasis. It included 38 lymph nodes in gastric samples, which achieved an accuracy of 96.33% overall. The diagnostic accuracy in this method was higher than the traditional ways of diagnostics, such as CT and multi-detector computed tomography (CT; 82.09%).

Support vector machines: SVM is a type of supervised learning used for both classification and regression problems. Using n as number of features, each data point is plotted in an n-dimensional space. As the name suggests, SVM transforms its data into the largest plane or support vectors of separation between two classes giving a flexible selection to have a wide plane.

This paper [47] describes a method using SVM to classify breast tissue using its multi-spectral MR images. The feasibility of this algorithm is evaluated by using likelihood ratios and classification rates, comparisons with the C-means, and a series of experiments and performance evaluations showing that SVM is an effective technique for the classification of MRI.

Decision tree: This algorithm is simple to understand and has human-readable rules. It results in a simple tree with the most possible and accurate outcomes. The algorithms allow us to choose the max depth and breadth for the search and have more decision points versus the right results. It gives a flowchart-like result where each point is an attribute following branch node theory.

The authors in [48] built and validated a classification and regression tree analysis to differentiate between lipoma and lipoma variants from liposarcoma by using an

MRI of the three with six findings such as shape, enhancement patterns, tumor size, anatomical location and was cross validated based on complexity parameter.

> *Naïve Bayes algorithm*: It is a collection of classification algorithms based on the traditional Bayes theorem, and every pair of classified features is independent of one another. It defines a relationship between the inputs and outputs.

Authors [49] developed a model to detect brain tissues affected by glioblastoma multiforme cancer, one of the growing brain tumors that quickly spreads to other parts of the brain. The naïve Bayes algorithm is used for recognizing the tumor, classifying it accurately, and detecting the regions with cancerous tissues. It detects tumor areas from various images and predicts the presence of tumors using statistical feature extraction, pre-processing MRI databases, and the naïve Bayes–based algorithm.

> *Artificial neural network*: This is one of the most famous algorithms used in AI that can solve complicated problems flexibly but is also complex and expensive to train. The neural network has several connection points called neurons which are organized in a layered-like structure. The data enters through the input layer, followed by many hidden layers where the data is transformed as it passes through it and finally, the output layer that makes the network's predictions. The network identifies patterns using the labeled data used for training and gives outputs that are compared to the actual labels. The network is trained, and every neuron is fine-tuned till the result gives out useful prediction. Once the network learns the data patterns, it can make predictions on new data.
>
> *Deep learning*: Also known as deep neural networks, it is a growing area of research in ML with the concept of artificial neural networks. It uses multiple layers of representation and is suited for complex problem-solving, making it applicable to multi-disciplinary fields. The unique part of DP is feature learning (i.e., layers learning the representation of data automatically). This makes deep neural networks differ from traditional ML methods [50]. Table 13.3 presents the DL contribution towards medical imaging.

13.6 OVERVIEW OF EDGE AI IN THE HEALTHCARE INDUSTRY

From the above discussion, we conclude that AI in healthcare can bring promising advances and improve accuracy and efficiency to a great level. Edge AI will bring ML/DL to the devices, creating more privacy and computational resource requirements and allowing offline execution. This section discusses how these combined technologies would work in a clinical workflow once applied in real-life scenarios.

From Table 13.3, we can confidently say that AI will become the new frontier in the healthcare industry. From receptionists and administrators using AI for scheduling, analyzing and preparing, technologists performing screening and detection, multiple physicians collaborating with the pathologist to diagnosis, reporting and

TABLE 13.3
Deep Learning Contribution towards Medical Imaging

Authors	Technique	Description	Application in Medical Imaging
[51]	GoogleNet combined with Bayesian Optimization algorithm and k-fold cross validation	A deep network of 22 layers whose quality is evaluated in terms of classification and detection. The Glomerulus Classification experiment indicated that this method performed better compared to other state-of-the-art methods.	Classification
[52]	Hybrid PCANet and DenseNet	A light weighted hybrid neural network that combines an unsupervised CNN (PCANet) and Supervised CNN (DenseNet) to achieve classification sensitivity and accuracy	Classification
[53]	Synergic DL model	ICH diagnostics utilizing GrabCUT-based segmentation with SDL (Synergy DL) termed GC-SDL to increase image quality and identify illness spots in the image using Gabor filtering for noise removal. The method achieved a specificity of 97.78%, precision of 95.79%, higher sensitivity of 94.01% and accuracy of 95.73%.	Classification and detection
[54]	ResNet-50 pretrained on ImageNet with supervised attention mechanism	MRI breast lesions detection and characterization. A lesion-characterization model was created based on a single two-dimensional T1-weighted fat-suppressed MR image produced following intravenous administration of a gadolinium chelate selected by radiologists. 335 MR scans from 335 patients were included in the study. A cross-validation score of 0.817 weighted average receivers operating characteristic (ROC)-AUC was obtained on the training set computed.	Detection
[55]	nnU-Net	This framework automatically adapts itself to any new dataset and successfully solves six segmentation challenges.	Segmentation
[56]	Self-ensembling semi-supervised model.	The model was tested on three approaches: 1. Skin lesion segmentation from dermoscopy images on the International Skin Imaging Collaboration 2017 dataset 2. Optic disc segmentation from fundus images on the Retinal Fundus Glaucoma Challenge (REFUGE) dataset 3. Liver segmentation from volumetric CT scans on the Liver Tumor Segmentation Challenge (LiTS) dataset It outperforms supervised algorithms on complex 2D/3D medical images, highlighting the utility of semi-supervised methods for medical image segmentation.	Segmentation

analyzing results to radiologists and physicians planning and executing the treatments with assessments and follow-up care. AI will be involved in every treatment step for patients to achieve better and more accurate results. AI does not replace physicians but helps them enhance their treatment and methodology, automating tasks and making an easy workflow environment. The figure below shows how every device in the hospital is connected from ambulances to emergency rooms to doctors and administrators in the front desk.

13.7   HOW 5G WILL IMPACT EDGE AI IN HEALTHCARE INDUSTRY

Latency is the measure of the delay in the communication network and can be considered one of AI's drawbacks as an immediate response is necessary, especially in the healthcare scenario. This can be compensated with the use of fifth-generation technology.

5G provides three key factors (Figure 13.4):

MBB (enhanced mobile broadband), which will provide gigabytes of bandwidth on our demand.

mMTC (massive machine type communication), which will connect billions of machines and devices.

URLLC (ultra-reliable low latency communication): These technologies can help healthcare massively by connecting billions of devices, communicating, and providing large bandwidths, especially for medical imaging and low latency that will reduce the delay in AI, prediction and analysis of results, thus making smart hospitals' workflow smooth and faster.

Combining 5G with Edge AI can help in in various domains, including:

Remote surgery: High transmission and real-time data transfer can be achieved by implementing 5G with Edge AI.

Connected ambulance: Faster and quicker services can be provided by paramedic staff to any patient who is critically ill if that ambulance is well connected with strong network such as 5G and enabled with Edge AI techniques.

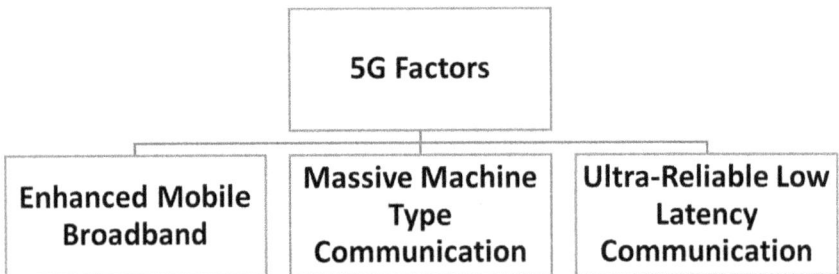

FIGURE 13.4   5G factors.

IoT Sensor Monitoring: In a wireless network, many sensors contribute to data collection and analysis. Monitoring of these sensors can also be done in real time with Edge AI techniques and 5G.

13.8 CHALLENGES OF EDGE AI AND MACHINE LEARNING IN HEALTHCARE INDUSTRY

Edge computing is decentralized, and having data pooled locally would demand security in more locations. These will increase the number of physical data points that make up the Edge AI architecture, making it more vulnerable to many cyber-attacks. Being AI to Edge means that the edge devices would require massive computational power to perform their tasks. Due to limited ML power, Edge AI can underperform and does not meet the requirements. In many cases, the large AI models will have to be simplified before deployment on the Edge AI hardware to meet accuracy and efficiency. Applying AI and ML in healthcare comes with many challenges:

The first is datasets. Finding the right dataset to train your model for a particular illness is critical. The datasets need to be huge for DL models to obtain accurate results. Not using proper or small datasets can result in models that may provide wrong insights, which can cause sensitive issues in healthcare.

The next challenge is finding the right expertise to train the model and get useful insights, especially in the healthcare sector; being able to decide what tools to use and what questions can be answered by the ML models, along with which model fits perfectly for the root of the analysis. Investments can be an issue as every healthcare organization may need help to build AI and equipment in their hospitals to utilize Edge AI. Together, these challenges can delay the implementation of Edge AI and ML in smart hospitals.

Furthermore, to make use of 5G technology in Edge AI, data privacy and network security will become a concern. Medical records are sensitive data, and to make patients feel safe without the fear of cyberattacks, the 5G network implemented in the healthcare system must be protected from cyber intruders at all costs.

13.9 CONCLUSION

AI has made its way into almost all industries in the world. AI in healthcare will revolutionize changes to the workflow system, and taking care of patients will become more efficient and accurate. Edge AI will bring AI models to Edge devices, making remote clinical work easier and faster; it will help in immediate analytics and communications between patients and doctors. AI can be applied to every aspect of the industry, from automating scheduling appointments to surgeries and predicting diseases, especially using medical imaging with DL. All these smart approaches will make what is called a smart hospital. Soon the implementation of 5G will help elevate communication and reduce latency, thus making decisions faster and providing more bandwidth for network connection. These smart approaches have proven to help the healthcare industry possible in many ways.

REFERENCES

[1]. Petiwala, F. F., Shukla, V. K., and Vyas, S. (2021). IBM Watson: Redefining artificial intelligence through cognitive computing. In: Prateek, M., Singh, T. P., Choudhury, T., Pandey, H. M., Gia Nhu, N. (eds) *Proceedings of International Conference on Machine Intelligence and Data Science Applications. Algorithms for Intelligent Systems.* Springer, Singapore. https://doi.org/10.1007/978-981-33-4087-9_15

[2]. Mathew, D., Shukla, V. K., Chaubey, A., and Dutta, S. (2021). "Artificial intelligence: hope for future or hype by intellectuals?," In: *2021 9th International Conference on Reliability, Infocom Technologies and Optimization (Trends and Future Directions) (ICRITO),* pp. 1–6, IEEE, Noida, India. doi:10.1109/ICRITO51393.2021.9596410.

[3]. Tamang, M. D., Kumar Shukla, V., Anwar, S., and Punhani, R. (2021). "Improving business intelligence through machine learning algorithms," In: *2021 2nd International Conference on Intelligent Engineering and Management (ICIEM),* pp. 63–68, doi:10.1109/ICIEM51511.2021.9445344.

[4]. Suhel, S. F., Shukla, V. K., Vyas, S., and Mishra, V. P. (2020). "Conversation to automation in banking through chatbot using artificial machine intelligence language," In: *2020 8th International Conference on Reliability, Infocom Technologies and Optimization (Trends and Future Directions) (ICRITO),* pp. 611–618, doi:10.1109/ICRITO48877.2020.9197825.

[5]. Trayush, T., Bathla, R., Saini, S., and Shukla, V. K. (2021). "IoT in healthcare: Challenges, benefits, applications, and opportunities," In: *2021 International Conference on Advance Computing and Innovative Technologies in Engineering (ICACITE),* pp. 107–111, doi:10.1109/ICACITE51222.2021.9404583.

[6]. Athota, L., Shukla, V. K., Pandey, N., and Rana, A. (2020). "Chatbot for healthcare system using artificial intelligence," In: *2020 8th International Conference on Reliability, Infocom Technologies and Optimization (Trends and Future Directions) (ICRITO),* pp. 619–622, doi:10.1109/ICRITO48877.2020.9197833.

[7]. Shukla, V. K., Wanganoo, L., and Tiwari, N. (2022). Real-time alert system for delivery operators through artificial intelligence in last-mile delivery. In: Garg, L., Chakraborty, C., Mahmoudi, S., and Sohmen, V. S. (eds) *Healthcare Informatics for Fighting COVID-19 and Future Epidemics. EAI/Springer Innovations in Communication and Computing.* Springer, Cham, doi:10.1007/978-3-030-72752-9_20

[8]. Bhardwaj, M., Pandey, N., Shukla, V. K., Singh, A. V., and Gupta, N. (2021). "Review and analysis of security model in healthcare system," In: *2021 9th International Conference on Reliability, Infocom Technologies and Optimization (Trends and Future Directions) (ICRITO),* pp. 1–7, doi:10.1109/ICRITO51393.2021.9596090.

[9]. Sanghar, M., Shukla, V. K., Verma, A., and Sharma, P. (2021). "Implementation of support vector machines algorithm through r-language for diabetes database testing," In: *2021 11th International Conference on Cloud Computing, Data Science & Engineering (Confluence),* pp. 746–751, doi:10.1109/Confluence51648.2021.9377124.

[10]. Andras, I., Mazzone, E., van Leeuwen, F. W., De Naeyer, G., van Oosterom, M. N., Beato, S., . . . and Mottrie, A. (2020). Artificial intelligence and robotics: A combination that is changing the operating room. *World Journal of Urology, 38*(10), 2359–2366.

[11]. Fard, M. J., Ameri, S., Darin Ellis, R., Chinnam, R. B., Pandya, A. K., and Klein, M. D. (2018). Automated robot-assisted surgical skill evaluation: Predictive analytics approach. *The International Journal of Medical Robotics and Computer Assisted Surgery, 14*(1), e1850.

[12]. Dai, Y., Abiri, A., Pensa, J., Liu, S., Paydar, O., Sohn, H., . . . and Candler, R. N. (2019). Biaxial sensing suture breakage warning system for robotic surgery. *Biomedical Microdevices, 21*(1), 1–6.

[13]. Hung, A. J., Chen, J., Che, Z., Nilanon, T., Jarc, A., Titus, M., . . . and Liu, Y. (2018). Utilizing machine learning and automated performance metrics to evaluate robot-assisted radical prostatectomy performance and predict outcomes. *Journal of Endourology*, *32*(5), 438–444.

[14]. Hung, A. J., Chen, J., Ghodoussipour, S., Oh, P. J., Liu, Z., Nguyen, J., . . . and Liu, Y. (2019). A deep-learning model using automated performance metrics and clinical features to predict urinary continence recovery after robot-assisted radical prostatectomy. *BJU International*, *124*(3), 487–495.

[15]. Zhao, B., Waterman, R. S., Urman, R. D., and Gabriel, R. A. (2019). A machine learning approach to predicting case duration for robot-assisted surgery. *Journal of Medical Systems*, *43*(2), 1–8.

[16]. Wang, Z., and Majewicz Fey, A. (2018). Deep learning with convolutional neural network for objective skill evaluation in robot-assisted surgery. *International Journal of Computer Assisted Radiology and Surgery*, *13*(12), 1959–1970.

[17]. Ershad, M., Rege, R., and Majewicz Fey, A. (2019). Automatic and near real-time stylistic behavior assessment in robotic surgery. *International Journal of Computer Assisted Radiology and Surgery*, *14*(4), 635–643.

[18]. Bhattacharya, A., Ghosh, G., Mandal, R., Ghatak, S., Samanta, D., Shukla, V. K., . . . and Mandal, A. (2022). Predictive Analysis of the Recovery Rate from Coronavirus (COVID-19). In: *Cyber Intelligence and Information Retrieval* (pp. 309–320). Springer, Singapore.

[19]. U. Thange, V. K. Shukla, R. Punhani, and W. Grobbelaar, "Analyzing COVID-19 Dataset through Data Mining Tool 'Orange,' " In: 2021 2nd International Conference on Computation, Automation and Knowledge Management (ICCAKM), 2021, pp. 198–203, doi:10.1109/ICCAKM50778.2021.9357754.

[20]. Latheef, A., Ali, M. F. L., Bhardwaj, A. B., and Shukla, V. K. (2021). Structuring Learning Analytics through Visual Media and Online Classrooms on Social Cognition during COVID-19 Pandemic. In: *Journal of Physics: Conference Series* (Vol. 1714, No. 1, p. 012019). IOP Publishing.

[21]. Sebastian, S., Thomson, A., Shukla, V. K., and Naje, I. A. (2022). Role of ICT in Online Education during COVID-19 Pandemic and beyond: Issues, Challenges, and Infrastructure. In: *ICT and Data Sciences* (pp. 249–276). CRC Press.

[22]. Java, S., Mohammed, H., Bhardwaj, A. B., and Shukla, V. K. (2021). Education 4.0 and Web 3.0 applications in enhancing learning management system: Post-lockdown analysis in covid-19 pandemic. *Knowledge Management and Web 3.0: Next Generation Business Models*, *2*, 85.

[23]. Ardabili, S. F., Mosavi, A., Ghamisi, P., Ferdinand, F., Varkonyi-Koczy, A. R., Reuter, U., and Atkinson, P. M. (2020). Covid-19 outbreak prediction with machine learning. *Algorithms*, *13*(10), 249.

[24]. Tapak, L., Hamidi, O., Fathian, M., and Karami, M. (2019). Comparative evaluation of time series models for predicting influenza outbreaks: Application of influenza-like illness data from sentinel sites of healthcare centers in Iran. *BMC Research Notes*, *12*(1), 1–6.

[25]. Koike, F., and Morimoto, N. (2018). Supervised forecasting of the range expansion of novel non-indigenous organisms: Alien pest organisms and the 2009 H1N1 flu pandemic. *Global Ecology and Biogeography*, *27*(8), 991–1000.

[26]. Anno, S., Hara, T., Kai, H., Lee, M. A., Chang, Y., Oyoshi, K., . . . and Tadono, T. (2019). Spatiotemporal dengue fever hotspots associated with climatic factors in Taiwan including outbreak predictions based on machine-learning. *Geospatial Health*, *14*(2).

[27]. Chenar, S. S., and Deng, Z. (2018). Development of artificial intelligence approach to forecasting oyster norovirus outbreaks along Gulf of Mexico coast. *Environment International*, *111*, 212–223.

[28]. Liang, R., Lu, Y., Qu, X., Su, Q., Li, C., Xia, S., . . . and Niu, B. (2020). Prediction for global African swine fever outbreaks based on a combination of random forest algorithms and meteorological data. *Transboundary and Emerging Diseases*, *67*(2), 935–946.

[29]. Raja, D. B., Mallol, R., Ting, C. Y., Kamaludin, F., Ahmad, R., Ismail, S., . . . and Sundram, B. M. (2019). Artificial intelligence model as predictor for dengue outbreaks. *Malaysian Journal of Public Health Medicine*, *19*(2), 103–108.

[30]. Leventakos, K., Helgeson, J., Mansfield, A., Deering, E., Schwecke, A., Adjei, A., . . . and Haddad, T. (2019). P1.16–14 effects of an artificial intelligence (AI) system on clinical trial enrollment in lung cancer. *Journal of Thoracic Oncology*, *14*(10), S592.

[31]. U. K., S., Sudhir, S., and Palaniappan, S. (2020). Elderly behavior prediction using a deep learning model in smart homes. In: Wason, R., Goyal, D., Jain, V., Balamurugan, S., and Baliyan, A. (ed) *Applications of Deep Learning and Big IoT on Personalized Healthcare Services* (pp. 115–131). IGI Global, doi:10.4018/978-1-7998-2101-4.ch008

[32]. Zaidi, S. F. N., Shukla, V. K., Mishra, V. P., and Singh, B. (2021). "Redefining home automation through voice recognition system," In: *Emerging Technologies in Data Mining and Information Security* (pp. 155–165). Springer.

[33]. S. Ibrahim, V. K. Shukla, and R. Bathla, (2020). "Security enhancement in smart home management through multimodal biometric and passcode." In: *2020 International Conference on Intelligent Engineering and Management (ICIEM)*, pp. 420–424, doi:10.1109/ICIEM48762.2020.9160331.

[34]. Khan, R., Shukla, V. K., Singh, B., and Vyas, S. (2021). "Mitigating security challenges in smart home management through smart lock," In: *Data Driven Approach Towards Disruptive Technologies* (pp. 61–71). Springer, Singapore.

[35]. Shukla, V. K., and Verma, A. (2019). "Model for user customization in wearable virtual reality devices with IoT for "low vision,"" In: *2019 Amity International Conference on Artificial Intelligence (AICAI)*, pp. 806–810, doi:10.1109/AICAI.2019.8701386.

[36]. Shukla, V. K., and Verma, A. (2019). "Enhancing user navigation experience, object identification and surface depth detection for "low vision" with Proposed electronic cane," In: *2019 Advances in Science and Engineering Technology International Conferences (ASET)*, 2019, pp. 1–5, doi:10.1109/ICASET.2019.8714213.

[37]. Monfared, M., Shukla, V. K., Dutta, S., & Chaubey, A. (2022). "Reshaping education through augmented reality and virtual reality," In: *Cyber Intelligence and Information Retrieval* (pp. 619–629). Springer, Singapore.

[38]. Terence, T., Grootboom, N., Zhou, M., Guvhu, R., and Shukla, V. K. (2021). "Addressing academic administration challenges in higher educational institutions through mobile technologies and cloud computing," In: *2021 9th International Conference on Reliability, Infocom Technologies and Optimization (Trends and Future Directions) (ICRITO)*, pp. 1–6, doi:10.1109/ICRITO51393.2021.9596144.

[39]. Zhou, S. K., Greenspan, H., Davatzikos, C., Duncan, J. S., Van Ginneken, B., Madabhushi, A., . . . and Summers, R. M. (2021). A review of deep learning in medical imaging: Imaging traits, technology trends, case studies with progress highlights, and future promises. *Proceedings of the IEEE*, *109*(5), 820–838.

[40]. Jha, S., and Topol, E. J. (2016). Adapting to artificial intelligence: Radiologists and pathologists as information specialists. *Jama*, *316*(22), 2353–2354.

[41]. Pesapane, F., Codari, M., and Sardanelli, F. (2018). Artificial intelligence in medical imaging: threat or opportunity? Radiologists again at the forefront of innovation in medicine. *European Radiology Experimental*, *2*(1), 1–10.

[42]. Golkov, V., Dosovitskiy, A., Sperl, J. I., Menzel, M. I., Czisch, M., Sämann, P., . . . and Cremers, D. (2016). Q-space deep learning: twelve-fold shorter and model-free diffusion MRI scans. *IEEE Transactions on Medical Imaging*, *35*(5), 1344–1351.

[43]. Lakhani, P., Prater, A. B., Hutson, R. K., Andriole, K. P., Dreyer, K. J., Morey, J., . . . and Hawkins, C. M. (2018). Machine learning in radiology: Applications beyond image interpretation. *Journal of the American College of Radiology*, *15*(2), 350–359.

[44]. Bharati, S., Podder, P., Mondal, M., and Prasath, V. B. (2021). Medical imaging with deep learning for COVID-19 diagnosis: A comprehensive review. *International Journal of Computer Information Systems and Industrial Management Applications*, *13*, ISSN 2150-7988, https://doi.org/10.48550/arXiv.2107.09602

[45]. Giger, M. L. (2018). Machine learning in medical imaging. *Journal of the American College of Radiology*, *15*(3), 512–520.

[46]. Li, C., Zhang, S., Zhang, H., Pang, L., Lam, K., Hui, C., and Zhang, S. (2012). Using the K-nearest neighbor algorithm for the classification of lymph node metastasis in gastric cancer. *Computational and Mathematical Methods in Medicine*, *2012*.

[47]. Lo, C. S., and Wang, C. M. (2012). Support vector machine for breast MR image classification. *Computers & Mathematics with Applications*, *64*(5), 1153–1162.

[48]. Shim, E. J., Yoon, M. A., Yoo, H. J., Chee, C. G., Lee, M. H., Lee, S. H., . . . and Shin, M. J. (2020). An MRI-based decision tree to distinguish lipomas and lipoma variants from well-differentiated liposarcoma of the extremity and superficial trunk: Classification and regression tree (CART) analysis. *European Journal of Radiology*, *127*, 109012.

[49]. Rajni, H., Soujanya, B., Varshitha, K. R., and Vinutha, S. M. (n.d.). *Brain tumor detection using naïve Bayes classification*. Retrieved February 2, 2022, from https://irjet.com/archives/V7/i6/IRJET-V7I61270.pdf

[50]. Lundervold, A. S., and Lundervold, A. (2019). An overview of deep learning in medical imaging focusing on MRI. *Zeitschrift für Medizinische Physik*, *29*(2), 102–127.

[51]. Yao, X., Wang, X., Karaca, Y., Xie, J., and Wang, S. (2020). Glomerulus classification via an improved GoogleNet. *IEEE Access*, *8*, 176916–176923.

[52]. Huang, Z., Zhu, X., Ding, M., and Zhang, X. (2020). Medical image classification using a light-weighted hybrid neural network based on PCANet and DenseNet. *IEEE Access*, *8*, 24697–24712.

[53]. Anupama, C. S. S., Sivaram, M., Lydia, E. L., Gupta, D., and Shankar, K. (2020). Synergic deep learning model–based automated detection and classification of brain intracranial hemorrhage images in wearable networks. *Personal and Ubiquitous Computing*, 1–10.

[54]. Herent, P., Schmauch, B., Jehanno, P., Dehaene, O., Saillard, C., Balleyguier, C., . . . and Jégou, S. (2019). Detection and characterization of MRI breast lesions using deep learning. *Diagnostic and Interventional Imaging*, *100*(4), 219–225.

[55]. Isensee, F., Petersen, J., Kohl, S. A., Jäger, P. F., and Maier-Hein, K. H. (2019). nnUnet: Breaking the spell on successful medical image segmentation. *arXiv preprint arXiv:1904.08128*, *1*(1–8), 2.

[56]. Li, X., Yu, L., Chen, H., Fu, C. W., Xing, L., and Heng, P. A. (2020). Transformation-consistent self-ensembling model for semisupervised medical image segmentation. *IEEE Transactions on Neural Networks and Learning Systems*, *32*(2), 523–534.

14 Factors Influencing the Building of Smart Cities in Developing Countries

The Case of Vietnam

Van Chien Nguyen and Lam Oanh Ha

CONTENTS

14.1 INTRODUCTION

The Fourth Industrial Revolution is a term first coined in Germany in 2011 as a foundation for creating economic development strategies based on the development of science and technology in countries. Under the development of the Internet era, the 4.0 technology revolution has completely changed human production activities. Via typical technology activities such as the Internet of Things (IoT), big data, the cloud and AI, data mining and augmented reality, and robotic process automation, the technological revolution focuses mainly on digital technology to provide comprehensively for production, boosting productivity, improving the quality of economic growth and changing people's lives, and many economic activities [1–3].

Facing the rapid change of the 4.0 technology revolution, many countries around the world have been building a smart city roadmap to enhance people's living quality. A smart city will apply the achievements of technology development in traffic management, analysis of socio-economic data for management and supervision, ensuring safety information security [1] [4], and monitoring security through a camera system. For the medical field, a smart city will be a place to connect the hospital

DOI: 10.1201/9781003218715-18

system, exchange patients and treatment information, deploy online public services, save time in healthcare and integrate with e-government. For education, smart city has solutions in training, online learning, data lookup, time and cost reduction. According to the assessment of Kosowatz [5], more and more people are choosing to live in cities (now half of the world's population), requiring cities to provide adequate services, essential services, entertainment needs, learning and utilities for the population. Kosowatz [5] evaluates that a smart city requires long-term and sustainable development. The advance of this smart city is closely tied to eight main areas: connectivity, care, healthcare, safety, water and energy, community interaction, housing and socio-economic development, and waste). Singapore and Dubai are seen as pioneering smart cities. In particular, Singapore has successfully become one of the smartest cities in the world. Today Singapore's success is contributed by the process of Singapore building a city with health programs, ensuring universal health, building a housing program for all, and the sharing of community values through connectivity through an eco-friendly transportation program, reducing traffic congestion and easily connecting, the public transport program is increasingly popular in Singapore [1].

Vietnam, a fast-growing country, is in Southeast Asia. From one of the poorest countries in the world, in 1986, Vietnam implemented economic renovation oriented towards a market economy. Vietnam has attained many achievements in socio-economic development; the average income will reach 7330 USD per person per year by 2023 (according to the exchange rates in May 2023), and the growth rate in the period of 34 years from 1986 to 2020 will reach over 10% on average. Along with the economic development, Vietnam's urban areas are also expanding; the urbanization rate has increased continuously in recent years. It increased rapidly from 19.6% in 2009 to 36.6% in 2016, with 629 cities and 820 cities, respectively. The urbanization rate by 2020 is about 40%, which is expected to increase to 40.91% in 2025 and 44.45% in 2030, with an urban population of about 47.25 million people in 2030 [2].

In the development program, Binh Duong is a locality allowed by the government to construct a smart city program as a foundation to promote other localities. On January 22, 2020, the smart city management board issued Plan No. 09/KH—BDH to conduct the Smart City project. The Vietnamese government selected Binh Duong because it is one of the top five provinces and cities attracting the largest foreign domestic investment (FDI) inflows in Vietnam and is also the locality with the highest per capita income in Vietnam [6]. According to the General Statistics Office, Binh Duong's per capita income is higher than Ho Chi Minh City and the capital Hanoi, the two largest cities in Vietnam. Therefore, this study will determine and evaluate the factors affecting the smart city construction process through a typical case study conducted in Binh Duong province, Vietnam. The research results will help researchers and governments of developing countries see the advantages and challenges in building smart with the Fourth Industrial Revolution and improving a high standard of living in each country.

Section 14.2 discusses the case study in Binh Duong and the orientation to build a smart city in Binh Duong. Section 14.3 discusses data and research methodology. Section 14.4 presents the results and section 14.5 the conclusion.

14.2 BINH DUONG AND THE TRENDS OF BUILDING SMART CITY

On January 1, 1997, Binh Duong province was established on the basis of separation from Song Be province. According to the Binhduong statistical yearbook (2019) [7], Binh Duong has nine district-level administrative units and 91 commune-level administrative units; the natural area is 2694.64 km²; the average population is 2,456,319 people (in which the labor force is 1,648,275 people with 52.74% male and 47.26% female). This is a province in the southeast region and is identified by the State of Vietnam as one of the provinces and cities in the Southern Key Economic Zone. Since its establishment, Binh Duong has oriented socio-economic development towards industrialization and modernization. Over 20 years of implementing this orientation, Binh Duong has had a sustainable economic structure with the proportion of industry and construction consistently accounting for more than 50% of the total GDP, while the average GDP per capita in 2019 reached 146.9 million VND, equivalent to 6344 USD (see Figure 14.1).

Figure 14.1 shows that the GDP of Binh Duong has increased continuously since 1996; however, since 2010, GDP has increased more strongly than in the previous period. For industry and construction, this sector's share is becoming more and more important to GDP, contributing from 45% of GDP in 1996 to over 66% of GDP in 2019. This result, which reflects local efforts to attract FDI through appropriate mechanisms and policies, based on building a government dedicated to serving investors, has led to an even greater scale of domestic and foreign investment capital in Binh Duong that is constantly increasing [8]. In particular, Binh Duong is highly appreciated by experts because it is a leading locality in developing the industrial park system in a new way. For example, Vietnam Singapore Industrial Park (VSIP I) Binh Duong was built in 1996 and is considered a model industrial park in Vietnam because of its focus on building infrastructure associated with urbanization [9].

As of December 31, 2019, Binh Duong had 29 industrial parks with a total area of 127.43 km². The rental rate reached 83.3%. At the same time, the province also

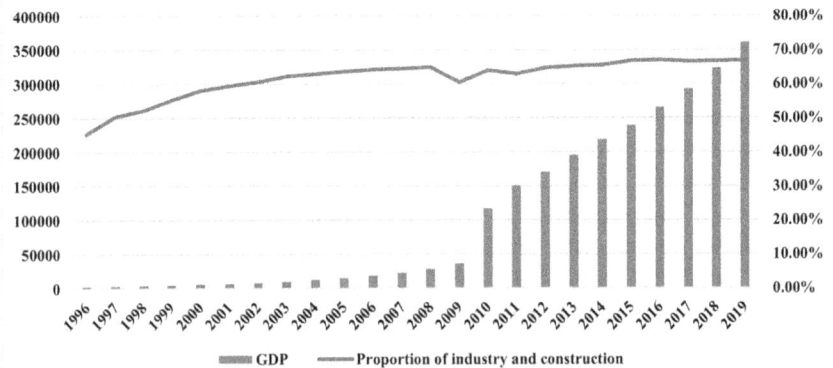

FIGURE 14.1 GDP size, share of industry and construction in Binh Duong province.

Source: Binhduong statistical yearbook.

developed 12 industrial clusters with a total area of 7.9 km^2 and an occupancy rate of 67.4% [7]. Over the years, some of the leading countries and territories in terms of investment capital in Binh Duong province are Japan, Korea, Singapore, Taiwan, Hong Kong, the United States, and the Netherlands. One thing these countries and territories have in common is that they lead the ranking of the smartest cities in the world [5].

Based on Decision No. 522/QD-BXD dated March 27, 2006, of the Ministry of Construction, Decision No. 310/QD-UBND dated January 31, 2008, and Decision No. 2273/QD-UBND dated June 2, 2009, of the People's Committee of Binh Duong province, on July 7, 2009, the People's Committee of Binh Duong province issued the decision No. 2717/QD-UBND on approving the grant of the license to use the right to the corporation Becamex IDC. This is considered the time to start implementing the Binh Duong New City construction project in a modern urban area. This project, with a scale of 10.11 km^2 located in six wards/communes (Phu My, Dinh Hoa, Phu Chanh, Tan Hiep, Tan Vinh Hiep, and Hoa Loi) in Thu Dau Mot city, Ben Cat town and Tan Uyen commune. In order to ensure the quality of the work, Binh Duong province has invited consultants from countries with the world's leading smart cities: the Design Research Institute of the National University of Singapore (NUS), CPG Company (Singapore), Kume Sekkei Company—KDA (Japan) and The Cox Group Company (Australia) [10]. According to the plan, Binh Duong new city includes the following main items:

- The political-administrative center will be the office of the Provincial Party Committee, the People's Council, the People's Committee and the departments and agencies of Binh Duong province. Next to the office block is a guest house and an international convention center.
- Eastern International University was built on an area of 0.26 km^2 and is a learning environment for about 24,000 students.
- MapleTree high-tech business town is invested by MapleTree Singapore with an area of 1.08 km^2, on which two-thirds will be built a high-tech park, with the rest offices and high-rise buildings. This is a technological knowledge city and a place to live and work for leading technical experts.
- The cultural center is a place for people to enjoy and exchange culture.
- Commercial, financial, banking, securities and office centers are ideal business locations for all investors.
- A living area with a variety of housing types suitable for people's income from ecological villas, townhouses and luxury apartments to social housing for low-income workers. The central park with two ecological lakes is the green lung for Binh Duong new city.

The internal traffic system is thoroughly planned and connected to the main arterial roads of the province, such as National Highway 13, DT741, DT742, DT743, DT746 and My Phuoc—Tan Van highway. In particular, the tram system is planned to connect from the central area to other locations of Binh Duong province and connect with the train system of Ho Chi Minh City. In addition, there are six new urban areas (VSIP II, Song Than 3, Dai Dang, Kim Huy, Dong An, and Phu An) that are allowed to be built according to the planning "Industrial—service—urban complex."

Through the cooperation with Eindhoven city government—Netherlands, Brainport Group—Netherlands and Becamex IDC Corporation, on November 21, 2016, Binh Duong Provincial People's Committee issued Decision No. 3206/QD-UBND, initially implemented as the Binh Duong smart city project. According to this project, Binh Duong's new city area is selected as the nucleus for deployment; it will spread to other localities. The project also identifies specific actions that will revolve around four areas: people, technology, businesses and foundational elements [11]. Moreover, in 2018, Binh Duong registered to become an Intelligent Communication Forum (ICF) member. This is the platform that helps Binh Duong establish relationships with smart communities around the world to exchange experiences, exchange information, and attract high-quality human resources to work in Binh Duong. It can be assessed that Binh Duong has the right orientations and appropriate smart city development roadmap. This is confirmed by three consecutive years (2019–2021) Binh Duong was selected as 1 out of 21 cities with typical smart city development strategies in the world. This is a list compiled by ICF (2021) and published annually [12].

14.3   DATA AND METHODOLOGY

14.3.1   DATA

This study is conducted through data collected through interviews with 15 experts who are directly involved, 15 experts are government leaders, CEOs of production and business companies, commercial banks, software providers, real estate agents, university leaders, IT professionals and students. In addition, the study also conducted a survey of 180 people living, studying and working in Binh Duong province.

14.3.2   METHODOLOGY

A qualitative method is used by the author through in-depth interviews with 15 experts with the following objectives:

*   Determine the factors affecting the construction of Binh Duong smart city.
*   Identify the benefits and disadvantages of building a smart city.
*   Help the author to adjust the question before conducting quantitative research.

Quantitative research was carried out using an online survey method via Google Docs to measure the influence of factors on the construction of Binh Duong smart city, conducted on 180 research samples.

14.4   RESULTS

14.4.1   QUALITATIVE ANALYSIS

To conduct qualitative research and conduct in-depth interviews, we conducted in-depth interviews with 15 experts who are government leaders, CEOs of production and business companies, commercial banks, software providers, real estate companies, university leaders, IT professionals and students.

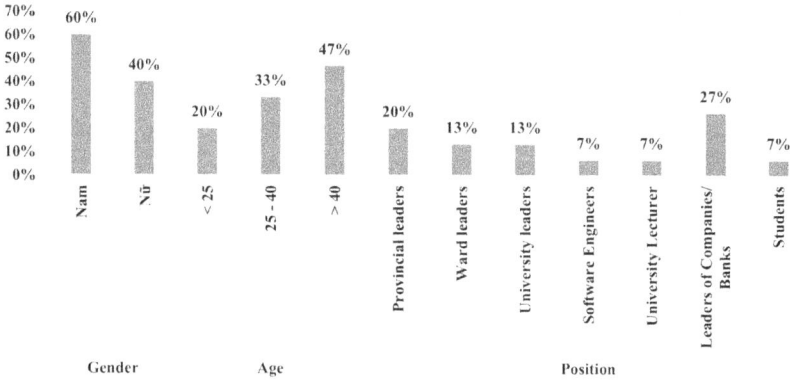

FIGURE 14.2 Characteristics of in-depth interview samples by gender, age and position.

Source: Analyzed by authors.

TABLE 14.1
Characteristics of In-Depth Interview Samples by Gender, Age and Position

	Criteria	Quantity	Percentage
Gender	Male	9	60%
	Female	6	40%
Age	Under 25	3	20%
	Between 25 and 40	5	33%
	Over 40	7	47%
Position	Director of the Department of Information and Communication	1	7%
	Director of the Department of Education	1	7%
	Director of the Department of Natural Resources and Environment	1	7%
	CEO of Homenext real estate company	1	7%
	Vice dean of Thu Dau Mot University	1	7%
	Deputy director of Institute of Technology and Technology—Thu Dau Mot University	1	7%
	Software engineer	1	7%
	Deputy director of MBBank Binh Duong	1	7%
	Chairman of the People's Committee of Phu Loi Ward	1	7%
	Chairman of the Women's Union of Phu Loi Ward	1	7%
	Lecturer in information technology	1	7%
	Director of a company providing enterprise management software	1	7%
	Garment company director	1	7%
	Banking and finance student	1	7%
	Electrical and electronic engineering student	1	7%

Source: Analyzed by authors.

Figure 14.2 and Table 14.1 show that, in the survey sample, the percentage of men and women interviewed is 60% and 40%, respectively, from many different professions and different positions in society, capable of understanding the smart city. The number of interviewees under age 25 years accounts for 20%, the rest 80% are experts who are over age 25 years. In addition, experts over age 40 account for the highest proportion, corresponding to 47% in the survey sample.

With the question "In your opinion, what is a smart city?", the interviewees all said that a smart city is a place where people's lives are humane, harmonious and balanced between the environment and life. Experts emphasize that smart cities must be environmentally friendly, and people use public transport. At the same time, this must be a place with high technology content and connect with other localities at home and abroad by a technology platform.

The experts' answers help the author confirm that there are four main factors affecting the construction of Binh Duong smart city. Experts also suggest the importance of these factors in descending order of magnitude, for example, government policy, people, technology and communication. The importance is ranked from 1 to 4, corresponding to increasing importance. The arguments they give to support this proposal are presented in Table 14.2.

Table 14.2 shows that through in-depth interviews with 15 experts, it was found that communication is an essential factor in the first stage of building a smart city. People's understanding of smart cities is still limited in the current stage, so communication becomes a good channel to help people understand smart cities, thereby attracting domestic and foreign investors, especially experts who come to Binh Duong to live. Table 14.2 also shows that the technology factor is considered to be very important in the process of building a smart city. Binh Duong in particular and Vietnam in general need practical solutions to attract technology. In addition, attracting high-quality human resources and appropriate government policies are factors that help build a smart city more effectively.

Besides the four main factors mentioned, the experts also discussed several other platforms contributing to the construction of Binh Duong smart cities, such as infrastructure, investment capital, and people's intellectual level. Further discussed in detail, experts all realize that Binh Duong already has these platforms since 2009 and deployed to build modern infrastructure compared to the common ground of Vietnam. At the same time, Binh Duong is a locality with a budget surplus, permanently doubling the budget expenditure (in 2019, budget revenues and expenditures were 70,284 VND and 31,436 billion VND, respectively). Accompanying the construction of a modern urban area is the birth of two universities (Thu Dau Mot University and Eastern International University) to improve the people's intellectual level and train human resources for service and local socio-economic development.

Assessing the benefits and disadvantages that smart cities bring to stakeholders, experts agree that in addition to the initial difficulties for the government when applying technology to provide public services, building a smart city will bring many benefits to the locality, especially in the long term. According to [13], smart city awareness is associated with a smart economy and governance. The obstacles of smart city construction often stem from regulations, work that requires a process, the

TABLE 14.2

Factors Affecting the Construction of Binh Duong Smart City

Factors	Experts' Comments	Importance Level (Number of Votes)			
		1	2	3	4
Government policies	1. This is a key factor. The government must be aware of building a smart city, creating a foundation for the government to decide to implement.	11	3	1	0
	2. Building a smart city is a long-term process, requiring the persistence of local authorities. About 5 years ago, people and other localities considered Binh Duong new city to be a "ghost city"; many thought that Binh Duong was wasteful when building a modern urban area dozens of kilometers away from the old center. If the government gives up, there will not be a new city in Binh Duong like today.				
	3. The succession of generations of government is required so that the deployment of smart city construction is not interrupted. In fact, in Binh Duong from 1997 until now, despite changing the leadership apparatus many times, the following leaders always inherit the achievements of the previous generation and develop in a better direction.				
Human capital	1. The smart city must be operated by smart people, so this is also a very important factor.	3	12	0	0
	2. That qualified human resources are able to communicate well in foreign languages and are capable of mastering technology are the most important factors.				
	3. Young intellectuals and their successors have the task of continuing to build and operate Binh Duong smart city, not only being good at their specialties but also knowing how to connect and associate with all related industries.				
Technology	1. A smart city is an environmentally friendly city that must have high technology content. A smart city that applies technological achievements will reduce the impact on the environment, promote the use of technology such as online shopping, online meetings, and using public transport.	1	2	12	0
	2. Technology is a tool to shorten work time, a foundation to increase the progress of smart city construction.				
Communication	1. Communication is a good channel for stakeholders to understand smart cities. Only when people correctly perceive the importance of building a smart city will the project succeed.	0	1	4	10
	2. Effective communication helps Binh Duong attract domestic and foreign investors and attract leading experts to live and work. From there, these subjects will contribute to perfecting the Binh Duong smart city ecosystem.				

Source: Analyzed by authors.

TABLE 14.3

Benefits from the Development of Binh Duong Smart City

Stakeholders	Benefits
Government	Building a civilized, advanced, mutually beneficial society.
	A good living environment will be a destination for domestic and foreign investors and talents.
Citizens	People can live in a civilized, safer environment, and their children will be well educated.
	They have the opportunity to expand relationships with external partners from which to gain more knowledge and increase income.
	Clean air helps people live healthy and long lives.
Financial institutions	A smart city with smart people helps to increase connectivity and improve the selling efficiency of financial institutions such as banks, securities companies and insurance companies.
Investors and firms	Saving resources from using common and modern infrastructure.
	Selecting quality workers, suitable for recruitment needs.
	Finding more business opportunities, connecting investments, selling more easily through e-commerce, digital marketing.
Labor	Learning and accessing modern knowledge associated with technology.
	Having the opportunity to access many job positions with better regimes and promotion opportunities.

Source: Analyzed by authors.

choice of stakeholders, and it takes time to solve. In the assessment of Binh Duong in building a smart city, outstanding benefits are presented in Table 14.3.

14.4.2 Quantitative Analysis

To conduct quantitative research, 180 residents of Binh Duong province were surveyed to measure the influence of factors on the construction of Binh Duong smart city. Of the 180 response samples, 93 survey participants were male (51.67%) and 87 female (48.33%). Most of the study sample is aged 26 to 35 years (98 people account for 54.44%). The majority of respondents were bankers (27 samples), university students (25 samples), teachers (23 samples), workers (22 samples) and software engineers (16 samples). Detailed information about the survey sample characteristics is shown in Table 14.4.

The survey questionnaire has 12 observed variables, of which there are three questions to measure government policy factors (GOV), three questions to measure human factors (HU), three questions to measure human factors (TECH), and three questions to measure the communication factor (MED). The survey table is designed according to the Likert scale with levels from 1 to 5, in which the average level is 3. The survey results are presented in Table 14.5. The average scores of the observed variables are

TABLE 14.4
Survey Sample Characteristics by Sex, Age and Occupation

Characteristics		Quantity	Percentage
Gender	Male	93	51.67%
	Female	87	48.33%
Age	Under 25	25	13.89%
	26–35	98	54.44%
	36–45	29	16.11%
	Over 45	28	15.56%
Position	Banker	27	15.00%
	Student	25	13.89%
	Teacher	23	12.78%
	Worker	22	12.22%
	Software engineer	16	8.89%
	Construction engineer	15	8.33%
	Real estate business	10	5.56%
	Self-employed	9	5.00%
	Public administrative officer	9	5.00%
	Doctor	4	2.22%
	Insurance business	4	2.22%
	Stockbroker	3	1.67%
	Layer	2	1.11%
	Others	11	6.11%

Source: Analyzed by authors.

TABLE 14.5
Descriptive Statistics of Variables

Code	Scale	Min	Max	Mean	Std. Dev.
GOV1	The government must have had the idea and awareness of building a smart city many years ago.	3	5	4.2722	0.65165
GOV2	The government must have actions and documents showing the determination to build a smart city.	2	5	4.1889	0.89173
GOV3	The government must implement the construction of a smart city.	2	5	4.2833	0.75129
HU1	People are the main subject to operate smart city.	2	5	4.0500	0.76638
HU2	Professionally qualified people, good at foreign languages and mastering technology, are fundamental factors in building a smart city.	1	5	4.2556	0.67316
HU3	People with a desire to live in a good environment are the driving force behind building a smart city.	2	5	4.2333	0.78472
TECH1	A smart city must have high technology application content.	1	5	4.2056	0.76933
TECH2	Technology is the tool for faster smart city deployment.	2	5	4.2500	0.74836
TECH3	Applying technology to life will reduce negative impacts on the environment, thereby contributing to perfecting a smart city.	2	5	4.1500	0.67134
MED1	It is necessary to communicate so that all classes of people understand what a smart city is.	1	5	3.7667	0.88096
MED2	It is necessary to communicate so that all classes of people agree to build a smart city.	2	5	4.0111	0.69500
MED3	The media will help attract domestic and foreign investors and talents to participate in smart city construction.	1	5	3.8944	0.67451

Source: Analyzed by authors.

all larger than the average of 3, of which 10 of 12 variables have an average score above 4. This shows that the respondents all believe that the factors proposed in the study have all contributed to building a smart city. The observed variable GOV3 has the highest average score (4.2833), showing that the government must act, build and make progress in the implementation of the project in order for people to believe it. This is also confirmed by the research of Nakano and Washizu [4]. Building a smart city will increase social capital through information technology because information technology will increase communication, equivalent to working in the community.

In order to specifically assess the level of agreement of survey participants, the study carried out detailed statistics on the number of votes that respondents corresponded to each measure in the Likert scale (1: Strongly disagree, 2: Disagree, 3: Neutral, 4: Agree and 5: Strongly agree). The results are presented in Table 14.6.

Table 14.6 shows that about 92.78% of respondents rate "Neutral" or higher, and 66.67% of respondents rate "Agree" and "Strongly agree." The rates of rating "Disagree" and "Strongly disagree" are very low (ranging from 0 to 5.56% for "Disagree" and from 0 to 1.67% for "Strongly disagree"). The survey results once again confirm that most of the survey respondents strongly agree that factors such as government policy, human resources, technology and communication impact building a successful smart city. These comments are similar to the study of Sivaramakrishnan [1] in Singapore. The author said that Singapore had built a cohesive community. Singapore also invests in

TABLE 14.6
Consent Grade of Survey Participants

Code	Likert Scale					Percentage (%)				
	1	2	3	4	5	1	2	3	4	5
GOV1	0	0	20	91	69	0.00	0.00	11.11	50.56	38.33
GOV2	0	8	33	56	83	0.00	4.44	18.33	31.11	46.11
GOV3	0	2	26	71	81	0.00	1.11	14.44	39.44	45.00
HU1	0	4	36	87	53	0.00	2.22	20.00	48.33	29.44
HU2	1	1	15	97	66	0.56	0.56	8.33	53.89	36.67
HU3	0	9	12	87	72	0.00	5.00	6.67	48.33	40.00
TECH1	1	5	17	90	67	0.56	2.78	9.44	50.00	37.22
TECH2	0	2	27	75	76	0.00	1.11	15.00	41.67	42.22
TECH3	0	1	26	98	55	0.00	0.56	14.44	54.44	30.56
MED1	3	10	47	86	34	1.67	5.56	26.11	47.78	18.89
MED2	0	1	39	97	43	0.00	0.56	21.67	53.89	23.89
MED3	1	1	42	108	28	0.56	0.56	23.33	60.00	15.56

Source: Analyzed by authors.

technology, creating convenience in transportation and communication, translation and traffic. In the study of Lim et al. [14], the authors believe that smart cities must be closely associated with technology and service platforms to serve people's transactions and production activities.

In addition, through the survey, several surveys also obtained some notable comments, such as the following:

- Binh Duong is a friendly locality. Binh Duong people live as beautiful as "Tchaikovsky's swans." If many other localities will increase the price of services on major holidays or special religious days, while the Binh Duong people have free food and water for tourists, this is a positive in attracting tourists, investors and talent.
- Binh Duong New City Park is large, green and cool, making our people feel like living in famous cities of Southeast Asia and the world such as Singapore, Amsterdam (the Netherlands) and Venice (Italy).
- About 10–15 years ago, I was very sceptical about the decision to build Binh Duong new city. However, about 5 years ago, I realized that the government had made the right decisions because the space of the new city created a modern feeling but was very peaceful and fresh.
- In the past, I wanted my children to go to Ho Chi Minh City to study at university. However, so far, I have always wanted to orient my children to study in Binh Duong. I believe that universities in Binh Duong province can train the most appropriate human resources for the locality.
- We believe entirely in the leadership of the Party and the government. All policies are consistent towards building a prosperous, strong, civilized and safe Binh Duong for all people.

14.5 CONCLUSION

Building a smart city is an inevitable development trend associated with the change of the Fourth Industrial Revolution. A smart city is a green (clean), beautiful, modern and friendly living environment. It is an ideal place for people to live a healthy and long life and a destination for talented people and investors domestically and internationally. Binh Duong is an industrial province in Vietnam that has been successful in attracting FDI and at the same time has the highest per capita income in the country in order to further improve people's living standards, so creating the smart city is inevitable.

The study carried out a quantitative study surveying 180 people living in the area, mainly bankers, university students, teachers, workers, software engineers and others (doctors, lawyers, stockbrokers, public administrative officers, insurance businessmen, construction engineers), and 15 experts who have positions in society in Binh Duong province. The results of the research show that communication and technology level are the most important factors for smart city construction in Binh Duong.

In addition, for achieving useful results, the quality of human resources and government policies are also important factors supporting the process of building a smart city. Therefore, in order to make steady progress in economic development in the long term, the authorities and people of Binh Duong must be the subjects to operate Binh Duong smart city. The province needs to improve effective communication, attract high technology investments, and prioritize technology-oriented start-up activities and sustainable development. In addition, Binh Duong needs to train high-quality human resources capable of applying technology and building a friendly government to continue supporting investment and promoting local economic development.

REFERENCES

[1]. Sivaramakrishnan, S. (2019). 3 reasons why Singapore is the smartest city in the world. Retrieved from www.weforum.org/agenda/2019/11/singapore-smart-city/

[2]. NCIF (2019). Urbanization trend in Vietnam in the period of 2021–2030 and some consequences. Retrieved from http://ncif.gov.vn/Pages/NewsDetail.aspx?newid=21873

[3]. Ha, L.O., Nguyen, V.C., Tien, D.D.T., & Ngoc, B.T.B. (2022). Factors affecting the intention of using fintech services in the context of combating of fake news. *Studies in Computational Intelligence*, 1001, 277–297.

[4]. Nakano, S., & Washizu, A. (2021). Will smart cities enhance the social capital of residents? The importance of smart neighborhood management. *Cities*, 115, 103244.

[5]. Kosowatz, J. (2020). Top 10 growing smart cities. Retrieved from www.asme.org/topics-resources/content/top-10-growing-smart-cities#:~:text=Top%2010%20Growing%20Smart%20Cities%201%20Singapore.%20Singapore%E2%80%99s,launched%20more%20than%2070%20smart%20city%20initiatives

[6]. Binhduong Statistics Office (2020). *Binhduong statistical yearbook 2019*, Statistical Publishing House.

[7]. Binhduong Statistics Office (2019). *Binhduong statistical yearbook 2018*, Statistical Publishing House.

[8]. Chau, Q. (2019). Binh Duong surpassing FDI attraction targets. Retrieved from www. vir.com.vn/binh-duong-surpassing-fdi-attraction-targets-71890.html

[9]. Becamex (2021). Hệ thống các khu công nghiệp VSIP. Retrieved from https://becamex. com.vn/du-an/he-thong-cac-khu-cong-nghiep-vsip/

[10]. Becamex Tokyu (2009). Binh Duong new city. Retrieved from www.becamex-tokyu. com/en/introduction_of_binh_duong_province/binh_duong_new_city/

[11]. Binh Duong people's committee (2018). Binh Duong smart city project: Part 1— Realizing breakthrough programs of Binh Duong province. Retrieved from www.bin- hduong.gov.vn/Lists/ThongTinTuyenTruyen/ChiTiet.aspx?ID=1451

[12]. ICF (2021). The Smart21 communities of the year. Retrieved from www. intelligentcommunity.org/smart21

[13]. Bjorner, T. (2021). The advantages of and barriers to being smart in a smart city: The perceptions of project managers within a smart city cluster project in Greater Copenha- gen. *Cities*, 114, 103187.

[14]. Lim, C., Cho, G.H., & Kim, J. (2021). Understanding the linkages of smart-city tech- nologies and applications: Key lessons from a text mining approach and a call for future research. *Technological Forecasting and Social Change*, 170, 120893.

Index

For Product Safety Concerns and Information please contact our EU
representative GPSR@taylorandfrancis.com
Taylor & Francis Verlag GmbH, Kaufingerstraße 24, 80331 München, Germany

www.ingramcontent.com/pod-product-compliance
Lightning Source LLC
Chambersburg PA
CBHW060403220326
41598CB00023B/3007

9 781032 111742